A PRIMER OF
DISCRETE MATHEMATICS

A PRIMER OF
DISCRETE MATHEMATICS

Daniel T. Finkbeiner II

Wendell D. Lindstrom

Kenyon College

W. H. Freeman and Company
New York

The quilt shown on the cover was appliqued and quilted by Mary Finkbeiner and her daughters, Susan, Heidi, Ann, and Kate. Photograph by Phil Samuell.

Library of Congress Cataloging-in-Publication Data

Finkbeiner, Daniel T. (Daniel Talbot), 1919–1986
  A primer of discrete mathematics.

  Bibliography: p.
  Includes index.
    1. Mathematics—1961–   . 2. Electronic data
processing—Mathematics. I. Lindstrom, Wendell D.
II. Title.
QA39.2.F53 1986   510   86-18349
ISBN 0-7167-1815-4

Printed in the United States of America

1 2 3 4 5 6 7 8 9 0 HL 5 4 3 2 1 0 8 9 8 7

Contents

Preface

During the three years in which this book was planned, written, tested in class, and evaluated, and then twice more revised, tested in class, and reevaluated, we often have been asked these two questions by interested colleagues whose specialties are other than mathematics or computer science: "What is discrete mathematics?" and "Why should it be offered as an introductory course?" Both questions are answered succinctly in the following excerpt from an editorial in *Science* by Lynn A. Steen, the president of the Mathematical Association of America.

> Traditional school mathematics aims to perfect arithmetic skills . . . as a basis for algebra, which itself becomes the cornerstone of trigonometry, analytic geometry and calculus. America was built with these tools, applied by generations of engineers and scientists to the physical environment in which we live. But today's growth industries are dominated by information, which is abstract and immaterial. Whereas the material world is modeled by calculus, the language of continuous change, the immaterial world of information requires discontinuous, discrete mathematics. Both genetic codes and computer codes are intrinsically discrete. Discrete mathematics basically deals with fancy ways of arranging and counting. It can be used to enumerate genetic patterns and to count the branches in computer algorithms; it can be used to analyze the treelike branching of arteries and nerves, as well as the cascading options in a succession of either-or decisions. It can tell us how many things are there as well as help us find what we want among a bewildering morass of possibilities.

At the start of this decade a few computer scientists emphasized the need for computer science students to become familiar with mathematical topics that were not included in the traditional mathematics curriculum, at least not until late in the undergraduate years. They urged the mathematics community to consider seriously the thesis that introductory calculus need not be the only port of entry to the undergraduate curriculum for students who might need a substantial background in mathematics to achieve their career goals. Although the proposal is still a lively source of controversy, experimental courses in discrete mathematics have been introduced recently in a number of colleges and universities to provide a new option in mathematics for well-prepared entering students. This book resulted from just such an experiment.

There is no dearth of published recommendations from professional organizations concerning the length, content, and emphasis of such a course, but there is a lack of agreement. Thus we think it would be useful to describe here the principles that we set for ourselves at the start of our experiment.

1. Although we planned to stay informed about suggested syllabi, we wanted to proceed independently, designing a course that we felt was appropriate for students in small to medium-sized four-year colleges.

2. In choosing among the many topics that can be classified as discrete mathematics, we wanted to select material that was lively, challenging, useful, and enjoyable. In particular, we wanted to pick topics that demonstrated some creative aspects of mathematics.

3. We wanted to make no assumptions about the future careers or current interests of students who enrolled in the course, and we wanted to make the course attractive to any student with the requisite preparation and interest. (We considered four years of strong performance in college preparatory mathematics with demonstrated algebraic skills as necessary preparation.)

4. We thought that the course should be structured to develop in its students the steady growth of that ephemeral quality of mind that we call mathematical maturity. (Hard to define but easy to recognize, mathematical maturity includes intellectual curiosity, patience, persistence, ingenuity, and probably much more.)

5. We also decided that this should be a one-semester course. Our decision was based on practical considerations concerning teaching resources at Kenyon and the desire to keep the project manageable. As it turned out, there is enough material in this book for a two-quarter sequence of courses that can be combined with a one-quarter course in applied linear algebra to produce a one-year course in discrete mathematics.

We now turn to a more detailed description of the content and organization of this book. There are three major subject areas — distinct but not independent.

Foundations: Chapters 1 and 2 concern basic concepts, notation, and methods that occur throughout much of contemporary mathematics: set theory, logic, theorems and proofs (direct, indirect, and inductive), the natural numbers, elementary counting techniques, binary relations and their representations by graphs and by matrices, special types of relations, functions, operations, and mathematical structures.

Combinatorics: Chapter 3 greatly extends the development of counting techniques begun in Chapter 1. By this point in the course, we have moved gradually but surely to a higher level of mathematical sophistication, and Sections 3.8 through 3.10 introduce and demonstrate two counting techniques that are more advanced than those previously considered: generating functions and recursion equations.

Graphs: Chapter 4 introduces graph theory mainly through the popular puzzles and problems that mark the history of its early development and serve well to motivate important concepts such as isomorphism, planarity, Euler trails and circuits, Hamilton paths and circuits, uses of colors in graphs, trees, and search algorithms. Chapter 5 develops algorithmic solutions of several contemporary applications of graph theory.

For reasons that vary from chapter to chapter we have organized each of the first three chapters into two parts:

In Chapter 1, "Informal Set Theory" (Sections 1.1 through 1.5) is concerned mostly with sets and counting; "Methods of Reasoning" (Sections 1.6 through 1.10) introduces the language and notation of logic and its application to mathematical reasoning. Thus the entire chapter provides a foundation that is essential for the rest of the course.

"Basic Concepts" (Sections 2.1 through 2.6) in Chapter 2 develops the concept of a binary relation and illustrates important types of binary relations. Section 2.7, "A Recapitulation" describes a mathematical structure, and may be deferred or omitted entirely, as the instructor chooses.

In Chapter 3, "Basic Methods of Counting" (Sections 3.1 through 3.7) demonstrates basic techniques of efficient counting, and "Further Counting Techniques" (Sections 3.8 through 3.10) introduces generating functions and recursion equations as examples of more powerful and sophisticated techniques of counting. "Further Counting Techniques" may be deferred or omitted.

Chapter 4 introduces graph theory, following a historical path of famous problems and their solutions.

Chapter 5 presents a few contemporary applications of graph theory and describes algorithms for their solution; this chapter may be deferred or omitted.

To summarize, in our opinion most first-level courses in discrete mathematics should include material comparable to that in Chapter 1, all but the final section of Chapter 2, the first seven sections of Chapter 3, and all of Chapter 4. Many semester-length courses will be able to include additional topics from Chapters 2, 3, and 5. These estimates are based upon three years of experience in teaching such courses at Kenyon during the period in which these class notes evolved into their present form.

In our opinion this book presents more than enough material to challenge a class of first-year students in many colleges and universities, giving those students a firm foundation for further study of mathematics, computer science, and a host of other important subjects. The optional material in Chapters 2, 3, and 5 can be used by the instructor to adjust the pace and content of the course according to the interests and ability of the class.

One section of an entry-level course in discrete mathematics has been offered at Kenyon in one semester of each of the academic years 1983–1984, 1984–1985, and 1985–1986, and will be continued in that manner according to present plans. Each time, the class contained students from all four class years, from undirected neophytes to able senior mathematics majors. We covered all but Sections 3.8 through 3.10 on the first trial, all but Chapter 5 on the second, and all but Section 2.7 and Chapter 5 on the third. Student course evaluations on each occasion urged convincingly that the pace of the course be reduced to provide time for more assimilation of new concepts and methods and for further exploration of ideas that attracted interest along the way. This advice is sound, we feel, and we plan to adjust the pace next year to set aside more time for reinforcement of ideas and techniques, and whatever else might increase the effectiveness and pleasure of learning. Ideally, each instructor should seek a pace that is best for the class. We have discussed this in some detail only to emphasize that correct pacing is more difficult to achieve in a course of this nature than in one that has been taught for many years.

Problem assignments present another pedagogical issue that has special importance for this course. The development of skill in analyzing and solving problems is a primary goal of any introductory course in discrete mathematics. For most of us the best way to learn mathematical methods is to practice regularly, applying those methods to specific exercises. But by its nature discrete mathematics is different from other introductory courses, such as calculus, that have well-organized underlying theories within which we learn special techniques for solving special problems. Lacking an integrated theory, discrete mathematics often requires its students to use ingenuity and insight in solving problems, which in many cases is a totally new and sometimes frustrating experience. Thus instructors should plan to use some class time regularly to answer questions, to invite students to propose solutions, and to listen carefully to what they say. Instructors might hear a solution that is so clever that they wish they had thought of it! They might also hear an argument that sounds correct but leads to results that are different from their own solution. During the process of resolving such an issue satisfactorily, both the class and the instructor learn a lot and enjoy doing so. Homework assignments should be read and evaluated with special care, with detailed comments written to help the student to acquire greater understanding.

Daniel T. Finkbeiner II
Wendell D. Lindstrom

A Postscript to the Preface

Acknowledgments had not yet been added to the preface at the time of the senior author's death, on March 28, 1986. Whatever Dan had wanted to add to the preface will have to be left unwritten.

First and foremost I wish to pay tribute to Dan. The idea of the book was his, and it was he who carefully planned and guided its growth from the very beginning to the time it was ready for the copy editor. I give credit to him for whatever I have learned about writing a textbook and, more important, for much of what I have learned about teaching mathematics.

Mary Finkbeiner has worked closely with the authors from the start. It was she who did the accurate typing and skillful artwork for the original manuscript. I express my personal and deepest gratitude to her for the invaluable work she has done to bring this book to its present form. She helped me all the way with all the details. As we worked we thought of Dan. Our thoughts of him are captured by the title (taken from an Elizabeth Barrett Browning sonnet) that he used for a talk on discrete mathematics: "How do I love thee? Let me count the ways."

It has been a pleasure to work with the people at W. H. Freeman and Company. I especially want to thank Jerry Lyons and Philip McCaffrey for their encouragement and expert guidance. Thanks also to Victor Klee, consulting editor in mathematics, for his useful and encouraging advice.

Suggestions made by those who have reviewed the manuscript have been most helpful. I thank Donald J. Albers, Menlo College; Underwood Dudley, DePauw University; Wayne Dymacek, Washington and Lee University; Howard Hiller, Columbia University; Glenn W. Hopkins,

University of Mississippi; and Francis Masat, Glassboro State College, for their reviews.

I am indebted to Kenyon College for a grant we received to help with the original version, and to Alice Straus for typing the answer section. To my wife, Miriam, I express appreciation for her many useful suggestions and her constant support. Finally, I thank those Kenyon students who have studied from the manuscript and made numerous contributions for improvement.

Wendell D. Lindstrom

A Note to
The Student

Problem solving, viewed broadly as the process of making rational decisions or of providing acceptable answers to nontrivial questions, is an inescapable daily activity. The quality of any decision is determined largely by the ability of the decision maker to understand the nature of the problem, to recognize the essential facts and assumptions that underlie the problem, to state the problem in clear and precise language, to use suitable modes of thought in seeking a solution, and sometimes to verify that a given solution is the best possible.

In this book we shall take a somewhat more restricted view of problem solving and confine ourselves mainly to questions that involve mathematics or logic; initially the mathematical or logical content of a question might be evident, or it might be so hidden that a mathematical solution does not seem possible. An ability to express ideas precisely and to reason carefully, as developed through practice in solving mathematical problems, is equally applicable in nonmathematical analysis and decision making.

Although we primarily intend that you have the experience of problem solving, we also hope that you will achieve the ability to

1. Acquire an improved understanding of the nature of mathematics as a creative activity of the mind.

2. Develop an appreciation for the power, beauty, and limitations of mathematical modes of thought and expression.

3. Enjoy mathematics as an intellectual activity and a lifelong recreation.

4. Broaden and deepen knowledge and skill in formal reasoning.

5. Master some basic techniques for representing discrete structures in mathematical form and for analyzing such structures.

6. Foster the habit of relevant inquiry — the practice, when confronted with a problem, of asking, "Does a solution exist? How many solutions? How are they related? How can I be sure? What would happen if I changed a particular aspect of the problem?" and so on.

You should be aware that these goals might not be achieved easily, and you should expect this course to be quite different from your previous mathematics courses, particularly if you have not yet studied mathematics at the college level. Some of the questions you will be asked to consider will be perfectly easy to understand but surprisingly difficult to answer. Plan to study mathematics regularly, every day if your schedule permits. Read the text carefully, paying special attention to terminology, notation, definitions, and theorems. Work the text examples with pencil and paper as well as reading them. Then start to solve the assigned exercises. Be prepared to find that some of them might not yield to your first efforts; if you seem to get nowhere on a particular exercise after 15 minutes or so of concentrated effort, set it aside and go on, returning to it some hours later after letting your subconscious mind work on it. If inspiration still eludes you, be sure to raise the question with your instructor, no later than during the next meeting of the class.

We hope that you will enjoy this book, be stimulated by it, and feel rewarded for your efforts in learning from it.

A PRIMER OF
DISCRETE MATHEMATICS

Chapter 1

FUNDAMENTAL CONCEPTS

The meaning of the word "primer" depends on whether it is pronounced with a short i, as in "swimmer," or with a long i, as in "time." A "short-i primer" refers to a small book of elementary principles of a subject, whereas a "long-i primer" is an agent that prepares a mechanism for its intended activity—as in priming a cannon, a deep-well pump, or perhaps a carburetor.

There should be no doubt that the short-i primer is the intended pronunciation of the title of this book. However, the meaning of the long-i primer describes our hope that this book will help to prepare readers for their intended activities. Those activities, whatever they might be, inevitably will include some form of *problem solving,* so it is appropriate that the principal objective of this book is to extend and develop the problem-solving skills of its readers.

Solving problems requires the ability to express ideas precisely and to reason carefully. Thus the first chapter is devoted to a study of informal set theory and logic, thereby laying the foundations that are essential for the understanding and application of mathematics of any kind. Chapter 1 includes a discussion of elementary counting techniques (to pave the way for the more sophisticated techniques in Chapter 3) and a discussion of the applications of logic to mathematical reasoning.

INFORMAL SET THEORY

1.1 Preliminary Remarks

Before beginning a systematic study of set theory and logic let us briefly discuss the meaning of discrete mathematics and the problem illustrated by the design on the cover of this book. *Discrete mathematics,* a relatively new term, refers to various mathematical techniques used by computer scientists and mathematicians in finite systems. A digital computer, for example, works with only a finite set of numbers, performs only a finite number of operations, and has only a finite set of memory locations; thus, it can be considered a finite machine. Without committing a gross oversimplification, we can consider discrete mathematics to be the study of mathematical properties of sets and systems that have only a finite number of elements.

Most of the topics considered in discrete mathematics are segments of larger areas of mathematics, such as set theory, logic, basic concepts, combinatorics, and graph theory, all of which have roots that are more than a century old. Recently there has been a resurgence of interest in these areas because of the availability of computers—computers can make the large number of calculations required in applications involving large finite sets or complex relationships within sets.

One particular problem in combinatorics is illustrated by the design on the cover. Generally speaking, when working combinatorial problems we are concerned with either counting large sets of objects or combining sets of objects to construct patterns of a prespecified form. The quilt shown on the cover is an example of the latter procedure. Mathematicians call this construction a *counterexample* because it also shows that a conjecture by Leonhard Euler, one of history's most famous mathematicians, was incorrect.

Near the end of his long and productive career, the Swiss mathematician, Leonhard Euler (pronounced "Oiler"), 1707–1783, became interested in what are now called *Graeco-Latin* squares. The cover photograph shows a square quilt that is an *orthogonal Graeco-Latin square of order ten*. Note that the quilt has 100 cells, each of which is a large colored square with a smaller colored square overlaid in the upper left corner.

A square *quilt* of order n is a pattern of n^2 congruent square cells, arranged in n horizontal rows and n vertical columns. The quilt is called a *Graeco square* of order n if and only if each cell is assigned one of n colors in such a way that each row and each column contains precisely one of the n colors. A *Latin square* is simply a distinguishing name for another Graeco square. Observe that the quilt pictured on the cover can be regarded as a Graeco square (the set of large cells) over which a Latin square (the set of small squares) has been superimposed, forming a pattern that is called a *Graeco-Latin square*. Finally, a Graeco square and a Latin square, each of order n, are said to be *orthogonal* if and only if their combination has the special properties that no two ordered pairs of a large square and a superimposed small square are colored alike, and each row and each column contains precisely one large square and one small square of each color.

Table 1.1-1 uses capital and lowercase letters instead of large and small squares in each cell to form an orthogonal Graeco-Latin square of order three. No two ordered pairs of a capital letter and a lowercase letter are alike, and each row and each column contains precisely one of the three capital letters and one of the three lowercase letters.

We now write a fixed positive whole number n in the form $n = 4k + r$ for some nonnegative whole number k and some whole number r between 0 and 3, inclusive. Euler proved that orthogonal pairs of Graeco-Latin squares of order n exist for all values of $n = 4k + r$ when $r \neq 2$ — that is, for all odd values of n and for all values of n that are divisible by 4. He also proved that such squares do not exist for $n = 2$, but he was unable to

Ac	*Bb*	*Ca*
Ba	*Cc*	*Ab*
Cb	*Aa*	*Bc*

Table 1.1-1

decide the question for $n = 6$. His famous conjecture can be stated as follows:

> No orthogonal pair of Graeco-Latin squares of order n exists for $n = 4k + 2$ for any positive whole number k.

No further progress was made on the problem until 1901, when G. Tarry, a French mathematician, proved that no orthogonal pair exists for $n = 6$; however, in 1958, the remaining conjecture was thoroughly shattered by the discoveries of R. C. Bose, S. S. Shrikhande, and E. T. Parker of the cover pattern for $n = 10$ together with a proof that pairs of orthogonal Graeco-Latin squares of order n exist for all $n = 4k + 2 > 6$.

Exercises 1.1

1. Show that Euler's conjecture is correct for $n = 2$. Make your reasoning clear and complete.

2. Write an orthogonal Graeco-Latin square of order three that is different from the example in Table 1.1-1.

3. Write an orthogonal Graeco-Latin square of order four.

4. Write an orthogonal Graeco-Latin square of order five.

5. Verify that the quilt shown on the cover is indeed an orthogonal Graeco-Latin square of order ten. Explain how you verified each of the required properties.

1.2   Sets, Elements, and Subsets

We regard a *set* S* as a collection or family of objects that we call the *elements* or *members* of *S*. Sets will be denoted by capital letters such as S, T, A, X, whereas the members of a set will be denoted by lowercase letters such as a, b, y, s. If x is a member of S, we write

$$x \in S,$$

---

* The study of sets began about a century ago with the work of Georg Cantor. (See the June 1983 issue of *Scientific American* for a sketch of Cantor's life with a summary of his controversial theories of sets.)

read as "x belongs to S," "x is an element of S," or "x is in S." To denote that x is not a member of S, we write

$$x \notin S.$$

Although the previous paragraph establishes language and notation, it does not define the words "set," "element," "member," or "belongs to." We accept these *undefined* terms as part of the basic language for our study of sets, and we understand their meaning intuitively.

Examples of Sets

1. The list of all persons enrolled in this course
2. The assortment of all items for sale in the Spring 1975 Sears catalog
3. The roster of all players for the New York Jets
4. The collection of articles on your desk
5. The family of all lines in a plane
6. The list of all Social Security numbers already assigned
7. The group of all possible license plate combinations of three letters followed by three digits

Let us consider the fourth listed example and suppose that there is nothing on your desk. In that case there are no members of the set; the set is *empty, void,* or *null,* and it is denoted by the special symbol $\varnothing$.

It is sometimes more convenient to describe a set symbolically by listing within one set of braces all of the elements of S or stating within braces one or more defining properties satisfied by each element of S but by no element not in S. For example, let N denote the set of all natural numbers, and let I denote the set of all integers; then we can write

$$N = \{0, 1, 2, 3, \ldots\},$$
$$I = \{0, 1, -1, 2, -2, 3, -3, \ldots\}.$$

Or if we assume that the set I is known, then we can write

$$N = \{x \in I \mid x \geq 0\}.$$

In words, N equals the set of all integers x such that x is greater than or equal to 0. The defining property notation consists of two statements enclosed by braces and separated by a vertical line that is read as "such that." The statement before the vertical line defines the type of element being considered (integers in this case), and the statement after the vertical line expresses the defining property (nonnegative) of the elements of that set.

For example, a positive integer x is said to be *prime* if and only if $x > 1$ and x is not evenly divisible by any positive integer except 1 and x. The set P of all primes can be expressed in set notation as follows:

$$P = \{x \in N \,|\, x > 1, \text{ and for } a,b \in N \text{ if } x = ab, \text{ then } x = a \text{ or } x = b\}.$$

Any positive integer that is neither 1 nor prime is said to be *composite*. The set C of all composites can be written

$$C = \{x \in I \,|\, x > 1 \text{ and } x \notin P\}.$$

Notice that the set

$$B = \{x \in N \,|\, x \text{ is even and } 10 < x^2 < 100\}$$

can be written more simply by listing

$$B = \{4, 6, 8\}.$$

The order in which the elements are written in the list is completely irrelevant. Hence, we could also write

$$B = \{8, 4, 6\} = \{6, 8, 4\}.$$

Normally when the elements of a set are listed, each element is listed only once. For example, the set C of the squares of all integers between -3 and 2, inclusive, can be written as

$$
\begin{aligned}
C &= \{x^2 \,|\, x \in I \text{ and } -3 \leq x \leq 2\} \\
&= \{9, 4, 1, 0, 1, 4\} \\
&= \{0, 1, 4, 9\}.
\end{aligned}
$$

There are times, however, when we need to take into account multiple occurrences of the same symbol. For example, if we want to know the number of letters in the word "occasions," we might be looking for the number of distinct letters of the alphabet that occur in that word (6), or we might be looking for the number of characters, ignoring duplication, required to write the word (9). A collection of symbols, some of which occur more than once, can be called a *multiset* whenever we need to make that distinction.

Two sets S and T are *equal* if and only if they have the same members;

that is, each member of S is a member of T, and each member of T is a member of S. Stated once more,

$$S = T \text{ means } "x \in S \text{ if and only if } x \in T."$$

Because the expression "if and only if" is often misunderstood, let us look carefully at the meaning of the preceding definition. The statement "$x \in S$ if and only if $x \in T$" means that

(1) $$x \in S \text{ if } x \in T,$$

and

(2) $$x \in S \text{ only if } x \in T.$$

Statement 1 means the same as

(1′) $$\text{if } x \in T \text{ then } x \in S,$$

whereas statement 2 means the same as

(2′) $$\text{if } x \in S \text{ then } x \in T.$$

Often the equivalence of statements 2 and 2′ is misunderstood, but the following example might help you to understand:

"Rain falls only if the sky is cloudy"

means the same as

"If rain falls, then the sky is cloudy."

To summarize, the statement

"$x \in S$ if and only if $x \in T$"

means the same as the statement

"If $x \in T$, then $x \in S$; and if $x \in S$, then $x \in T$."

If each member of a set S is also a member of T, we say that S is a *subset* of T and write it as

$$S \subseteq T.$$

This notation is analogous to the notation for numerical inequality, $m \leq n$, which can also be written as $n \geq m$, so we also use the notation

$$T \supseteq S$$

to denote that T has S as a subset.

Using the definitions of equality and subset, we can describe the equality of sets by the subset relation

$$S = T \quad \text{if and only if} \quad S \subseteq T \text{ and } T \subseteq S.$$

When $S \subseteq T$ but $S \neq T$, we say that S is a *proper* subset of T, denoted by $S \subset T$. Observe that every set is a subset of itself:

$$S \subseteq S \qquad \text{for every set } S.$$

Also observe that the void set $\varnothing$ is a subset of every set:

$$\varnothing \subseteq S \qquad \text{for every set } S.$$

We can verify this assertion by asking whether there is a member of $\varnothing$ that is not a member of S. Because $\varnothing$ has no members, the answer is no, so any member of $\varnothing$ is a member of S.

It is now easy to deduce that there is only one empty set; the set of apples in a certain empty barrel is the same as the set of squirrels in a certain empty cage. To prove this, let $\varnothing(a)$ denote the set of apples in the empty barrel and $\varnothing(s)$ the set of squirrels in the empty cage. Because an empty set is a subset of every set, we have

$$\varnothing(a) \subseteq \varnothing(s) \qquad \text{and} \qquad \varnothing(s) \subseteq \varnothing(a),$$

so $\varnothing(a) = \varnothing(s)$. This justifies our use of the single symbol $\varnothing$ for the one empty set.

EXAMPLE 1   It is important to observe the distinction between an element of a set and the set itself. Thus the set consisting of one specific apple must be distinguished from that apple. An apple is an object that can be eaten, but the concept of the set whose only member is an apple exists only in our imagination.   ■

EXAMPLE 2 Let A denote the set of letters in the word "apple," B the set of letters in the word "plea," C the set of the letters in the word "lap," and D the set of letters in the word "alps." Then

$$A = B, \quad C \subset A, \quad C \subset B, \quad C \subset D.$$

Observe, however, that neither A nor D is a subset of the other. ∎

EXAMPLE 3 What is meant by the notation $\{\varnothing\}$? The braces signify a set, and the members are listed within the braces. So $\{\varnothing\}$ denotes the set whose only member is the empty set;

$$\varnothing \in \{\varnothing\}.$$

What is $\{\varnothing, \{\varnothing\}\}$? It is a set whose only members are the empty set $\varnothing$ and the set $\{\varnothing\}$ whose only member is the empty set. Thus $\varnothing$ has zero members, $\{\varnothing\}$ has one member, and $\{\varnothing, \{\varnothing\}\}$ has two members. Similarly $\{\varnothing, \{\varnothing\}, \{\{\varnothing\}\}\}$ has three members, the new member being the set whose only member is the set whose only member is the empty set. Although we shall not pursue the notion, this example indicates a method by which the empty set can be used to construct the set N of natural numbers. ∎

Relations between sets can be illustrated by a schematic drawing, called a *Venn diagram,* in which each set is represented as the interior of a closed region of the plane. The Venn diagram shown in Figure 1.2-1 illustrates that set A is a subset of set B because every point within A also lies within B.

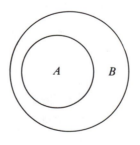

Figure 1.2-1

EXAMPLE 4   A small girl goes out to play with her friends, carrying a penny, a nickel, and a dime in her pocket. She does not lose or spend any coins that day, but because she is a generous child she might give some or all of her funds to her friends.

>   **(a)** How many different amounts of money might she have at the end of the day? List each of those amounts.
>
>   **(b)** Suppose the girl also found a quarter on her way to meet her friends. How many different amounts of money might she have at the end of the day?

To answer **a** we consider the various combinations of coins that the child might bring home. Each is a subset of the set

$$C = \{1, 5, 10\}.$$

A moment's reflection should convince us that different subsets of C cannot produce the same amount of money, therefore we must list and count all subsets of C. The subsets are

$$\emptyset, \{1\}, \{5\}, \{10\}, \{1, 5\}, \{1, 10\}, \{5, 10\}, \{1, 5, 10\}.$$

Hence, the number of cents the girl has when she gets home is one of the following:

$$0, 1, 5, 10, 6, 11, 15, 16. \qquad \blacksquare$$

EXAMPLE 5   This example illustrates that an intuitive study of set theory can lead to perplexing questions. Suppose we let $\mathcal{U}$ denote the set of all sets; that is, every set is a member of the set $\mathcal{U}$. Then $\mathcal{U}$ is a member of itself, $\mathcal{U} \in \mathcal{U}$, because $\mathcal{U}$ is a set. Thus it appears that there are sets that are members of themselves. Because this seems strange, let us call such sets "weird sets."
   Now let S be the subset of $\mathcal{U}$ defined by

$$S = \{X \in \mathcal{U} \mid X \text{ is not weird}\}.$$

As an exercise, show the following:

>   **(a)** If S is weird, then S is not weird.
>
>   **(b)** If S is not weird, then S is weird.

From statements **a** and **b** we realize that it is impossible to answer the question, Is the set S weird or not weird? For after all, each assumption leads us to a contradiction. This example is one form of the Russell paradox, named after Bertrand Russell, the English mathematician and philosopher.   ∎

Exercises 1.2

1. Describe in words each of the following sets.

   (a) $\{x \in N \mid x$ is evenly divisible by 3, and $15 < x\}$.
   (b) $\{x \in I \mid x^2 - x - 6 = 0\}$.
   (c) $\{3, 4, \{1, 2\}\}$.

2. Answer the following questions.

   (a) For each of the sets in Exercise 1, define a subset consisting of exactly two elements. Give your answers in set notation. Which of these are proper subsets?
   (b) List all the subsets of the set given in Exercise 1c.

3. Describe in set notation each of the following sets:

   (a) The set of all possible sums when a pair of dice is tossed.
   (b) The set of all natural numbers that are powers of 2.

4. Which of the following are correct assertions? Which are incorrect? Explain your answers.

   (a) $4 \in \{3, 4, 10\}$.
   (b) $\{4\} \in \{3, 4, 10\}$.
   (c) $\{4\} \subseteq \{3, 4, 10\}$.
   (d) $5 \notin \{3, 4, 10\}$.

5. List all the subset relations among the following five sets.

$$A = \{x \in I \mid x^2 \le 9\},$$
$$B = \{-2, 0, 2\},$$
$$C = \{-2, 0, 2, 4\},$$
$$D = \{x \in N \mid x^2 < 0\},$$
$$E = \{-3, -2, -1, 0, 1, 2, 3\}.$$

6. Let $S = \{1, \{1\}\}$. Which of the following are correct assertions? Which are incorrect? Explain your answers.

   (a) $1 \in S$.
   (b) $\{1\} \subseteq S$.
   (c) $\{1\} \in S$.
   (d) $\{1, \{1\}\} \in S$.

7. Answer Example 4b.

8. Answer Example 4b, but with this change: Instead of finding a quarter, the girl finds a buffalo nickel. (Because the two nickels can be distinguished from one another, you should think of the set C of coins as containing four distinct elements.)

9. Refer to Example 5. Explain why statements **a** and **b** are valid.

10. A certain student, M, claims to hand in the homework papers of all those students, and only those students, who do not hand in their own papers. Try to answer the following question and give the reasons for your response: Who hands in M's own homework papers?

1.3   An Algebra of Sets

Because the membership list of every human organization (political, social, religious, educational, commercial, philanthropic, scientific, musical, and so on) can be defined as a set it is useful to have a language and a system of notation to designate the common ways of combining sets to obtain new sets. This section introduces the basic features of such a language and its associated notation.

Let S denote a set, and let A and B be subsets of S. One method by which A and B can be combined to create another set is to put all the elements of A and all the elements of B into a single set, called the *union* of A and B, denoted $A \cup B$, and read as "A union B."

Set Union $A \cup B = \{x \in S \mid x \in A \text{ or } x \in B\}$.

Here the word "or" is used in the inclusive sense: x is a member of A or of B or of both; equivalently, x is a member of *at least one* of A and B. A second method of combining A and B is to select all the elements of S that belong to both A and B; the resulting set is called the *intersection* of A and B, denoted $A \cap B$, and read as "A intersect B."

Set Intersection $A \cap B = \{x \in S \mid x \in A \text{ and } x \in B\}$.

Venn diagrams illustrate these definitions. See Figure 1.3-1.

Of course, any diagram can represent only one of the possible ways that A and B are related. Conceivably A and B have *no* members in common.

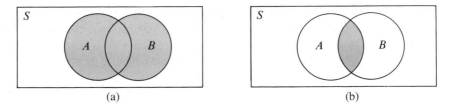

Figure 1.3-1 (a) $A \cup B$ shaded; (b) $A \cap B$ shaded.

Then $A \cap B = \varnothing$, and we say that A and B are *disjoint*. Or perhaps A is a subset of B, in which case $A \cup B = B$ and $A \cap B = A$. (Be sure you understand why.)

EXAMPLE 1   Let $S = \{0, 1, 2, \ldots, 9\}, A = \{2, 6, 7\}, B = \{0, 2, 4, 6, 8\}, C = \{1, 3, 5, 7, 9\}$. Then

$$A \cup B = \{0, 2, 4, 6, 7, 8\}, \qquad A \cap B = \{2, 6\},$$
$$A \cup C = \{1, 2, 3, 5, 6, 7, 9\}, \qquad A \cap C = \{7\},$$
$$B \cup C = S, \qquad B \cap C = \varnothing. \qquad \blacksquare$$

Because both union and intersection combine any two subsets of S to form a subset of S, each is called a *binary operation* on the family of all subsets of S. These operations on sets are roughly analogous to the operations of addition and multiplication of numbers, but we shall now see that the algebra of subsets is different in some respects from the algebra of numbers.

Theorem 1.3-1   The following subset relations hold for *all* subsets A, B, C of S.

1. $A \subseteq A$.
2. If $A \subseteq B$ and $B \subseteq A$, then $A = B$; and if $A = B$, then $A \subseteq B$ and $B \subseteq A$.
3. If $A \subseteq B$ and $B \subseteq C$, then $A \subseteq C$.
4. $A \subseteq A \cup B$.
4*. $A \supseteq A \cap B$.
5. If $A \subseteq B$, then $A \cup B = B$.
5*. If $A \supseteq B$, then $A \cap B = B$.
6. If $A \cup B = B$, then $A \subseteq B$.
6*. If $A \cap B = B$, then $A \supseteq B$.

Each of these properties is an immediate consequence of the definitions of subset, set equality, union, and intersection. Property 6 is called the *converse* of property 5, which means that the hypothesis of property 6 is the conclusion of property 5 and the conclusion of property 6 is the hypothesis of property 5. Thus the second part of property 2 is the converse of the first part. (The asterisk that follows the number of each of the last three properties has a significance that will be explained later in this section.)

Theorem 1.3-2   The following equations hold for *all* subsets A, B, C of S.

7. $A \cup B = B \cup A.$	Union is *commutative*.
7*. $A \cap B = B \cap A.$	Intersection is *commutative*.
8. $A \cup (B \cup C) = (A \cup B) \cup C.$	Union is *associative*.
8*. $A \cap (B \cap C) = (A \cap B) \cap C.$	Intersection is *associative*.
9. $A \cup (B \cap C) = (A \cup B) \cap (A \cup C).$	Union is *distributive* over intersection.
9*. $A \cap (B \cup C) = (A \cap B) \cup (A \cap C).$	Intersection is *distributive* over union.
10. $A \cup A = A.$	Union is *idempotent*.
10*. $A \cap A = A.$	Intersection is *idempotent*.
11. $A \cup \varnothing = A.$	$\varnothing$ is an *identity* relative to union.
11*. $A \cap S = A.$	S is an *identity* relative to intersection.

This theorem can be summarized by stating that set union and set intersection are both commutative and associative binary operations, each distributes over the other, each is idempotent, and each contains an identity element. If we compare these algebraic properties with corresponding properties of numerical algebra, we see that the analogy is strong but not perfect. Numerical addition and multiplication are *commutative* because

$$a + b = b + a \quad \text{and} \quad ab = ba \qquad \text{for all numbers } a, b.$$

Both are *associative* because

$$a + (b + c) = (a + b) + c \quad \text{and} \quad a(bc) = (ab)c \quad \text{for all numbers } a, b, c.$$

Zero is an *identity* of addition, and 1 is an *identity* of multiplication because

$$a + 0 = a \quad \text{and} \quad a(1) = a \qquad \text{for all } a.$$

Multiplication is commutative and associative because

$$ab = ba, \quad a(bc) = (ab)c \qquad \text{for all numbers } a, b, c.$$

Multiplication *distributes* over addition because

$$a(b + c) = ab + ac \qquad \text{for all numbers } a, b, c.$$

But this is as far as the analogy holds. It appears that union corresponds to addition, intersection corresponds to multiplication, $\varnothing$ corresponds to 0, and S corresponds to 1. But numerical addition does not distribute over multiplication because $a + bc$ does not equal $(a + b)(a + c)$ for all numbers; for example,

$$4 + 2(3) \neq (4 + 2)(4 + 3).$$

However, union distributes over intersection, and intersection distributes over union. Furthermore, neither idempotent property holds for numbers because

$$x + x = x \quad \text{if and only if} \quad x = 0,$$
$$x(x) = x \quad \text{if and only if} \quad x = 0 \text{ or } x = 1.$$

Here is a unary operation on the subsets of S, one that has no analog in numerical algebra. Given any subset A of S, the *complement* A' of A relative to S is defined as the set of all members of S that are not members of A. Thus an element x of S is a member of A' if and only if x is *not* a member of A.

Set Complement   If $A \subseteq S$, $A' = \{x \in S \mid x \notin A\}$.

Figure 1.3-2 is a Venn diagram of the complement of a set.

Theorem 1.3-3   The following relations hold for *all* subsets A, B of S.

   12. $(A')' = A$.
   13. If $A \subseteq B$, then $B' \subseteq A'$, and conversely.
   14. $A \cup A' = S$.
   14*. $A \cap A' = \emptyset$.
   15. $(A \cup B)' = A' \cap B'$.
   15*. $(A \cap B)' = A' \cup B'$.

   Each of the first four properties follows directly from the definition of complement. The last two equations are called the *De Morgan laws.* In words, property 15 says "the complement of the union of two sets is the intersection of the two complements." Property 15* reads "the complement of the intersection of two sets is the union of the two complements."
   Look carefully at each property marked with an asterisk and compare it with its numerical twin without an asterisk. There are 10 such pairs in the three theorems. Notice that if you take any one of these 20 statements, its paired associate can be obtained by replacing

   1. each $\cup$ by $\cap$ and each $\cap$ by $\cup$,
   2. $\emptyset$ by S and S by $\emptyset$,
   3. each $\subseteq$ by $\supseteq$ and each $\supseteq$ by $\subseteq$.

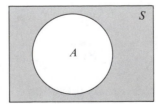

Figure 1.3-2 A' shaded.

This transformation is known as *duality* in the algebra of sets. The paired statements are called *dual* statements. The *principle of duality* assures us that whenever a statement about set union, intersection, or complementation is true for all sets, the dual statement is also true for all sets.

To illustrate how these theorems can be proved, Example 2 provides a method of proof for property 15.

EXAMPLE 2   Prove that $(A \cup B)' = A' \cap B'$.

Let A, B be subsets of S. To prove that the two sets are equal, we show that each element of $(A \cup B)'$ is an element of $A' \cap B'$, and conversely. Let $x \in (A \cup B)'$. Then $x \notin A \cup B$. Then $x \notin A$ and also $x \notin B$. So $x \in A'$ and $x \in B'$, establishing that $x \in A' \cap B'$. Conversely, let $y \in A' \cap B'$. Then $y \in A'$ and $y \in B'$ (by definition of intersection). Thus $y \notin A$ and $y \notin B$ (by definition of complement). But if y is in neither A nor B, y is not in $A \cup B$. That is, $y \in (A \cup B)'$ (by definition of complement).   ∎

Finally, pay special attention to the three small words "or," "and," and "not" and the vital role that they have in the basic definitions of the algebra of sets. We shall observe this important role again in Section 1.6.

Exercises 1.3

1. Let $A = \{2, 3, 4, 5, 6\}$, $B = \{1, 6, 7\}$, and $C = \{x \in N \,|\, x^2 < 9\}$.

   (a) Determine the members of the sets $B \cap C$, $A \cup (B \cap C)$, $A \cup B$, $A \cup C$, and $(A \cup B) \cap (A \cup C)$. (Note that the second and fifth sets have the same membership lists, which illustrates Equation 9 of Theorem 1.3-2 for the sets A, B, C given in this exercise.)

   (b) As in part **a**, verify Equation 9* of Theorem 1.3-2 for the sets A, B, C.

2. Let R and T be subsets of a set W.

   (a) Draw a Venn diagram that represents the set $(R \cap T)'$. Do the same for the set $R' \cup T'$. If these two diagrams are identical, this provides strong evidence that Equation 15* of Theorem 1.3-3 is valid.

   (b) Now prove Equation 15* of Theorem 1.3-3 by using the method of proof used in Example 2.

3. Use Theorems 1.3-1 and 1.3-2 to simplify the following expressions.

   (a) $A \cup (A \cap B)$.
   (b) $A \cap (A \cup B)$.

4. The identities given in the three theorems of this section can be used to simplify expressions involving sets. As an illustration, let A, B, C be subsets of S; then we have

$$
\begin{aligned}
(A \cap B) \cup (A' \cup B)' &= (A \cap B) \cup ([A']' \cap B') \\
&= (A \cap B) \cup (A \cap B') \\
&= [(A \cap B) \cup A] \cap [(A \cap B) \cup B'] \\
&= A \cap [(A \cap B) \cup B'] \\
&= A \cap [(A \cup B') \cap (B \cup B')] \\
&= A \cap [(A \cup B') \cap S] \\
&= A \cap (A \cup B') \\
&= (A \cap A) \cup (A \cap B') \\
&= A \cup (A \cap B') \\
&= A.
\end{aligned}
$$

(a) Identify the identities used in each step of the illustration. Use the identities in the theorems of this section to show that each of the following equations is an identity.

(b) $(A \cap B')' \cup B = A' \cup B$.

(c) $[(A \cap B') \cup (A \cap B)] \cup C = A \cup C$.

(d) $A \cap (B \cap C') = (A \cap B) \cap (A \cap C)'$.

(e) $[(A \cap B') \cup (A' \cap B)] \cap (A \cap B) = (A \cap B') \cup (A' \cap B)$.

5. Illustrate the following equation by using Venn diagrams.

$$
(A \cap B') \cap C' = A \cap (B \cup C)'.
$$

Begin with the diagram

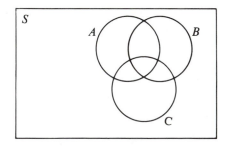

then sketch separate diagrams for the sets $A \cap B'$ and C' and one for $(A \cap B') \cap C'$. Repeat this procedure to obtain a diagram for the right-hand set, $A \cap (B \cup C)'$, and compare the two results.

6. Given two sets A, B, the *difference* of A and B, denoted by $A - B$, is defined by

$$A - B = A \cap B'.$$

Note that $A - B$ is simply the set of all members of A that are *not* members of B.

(a) Show that the equation in Exercise 5 may be written as

$$(A - B) - C = A - (B \cup C).$$

(b) Use Venn diagrams to illustrate the equation

$$A - (B - C) = (A - B) \cup (A \cap C).$$

(c) Is set difference an associative operation? Justify your answer.

7. Let A denote the set {d, g, i, t} of letters in the word "digit," B the set of letters in the word "numeral," and C the set of letters in the word "iteration."

(a) Determine the numbers of elements in the following sets: $A \cap B$, $A \cap C, A \cup B, A \cup C$.

(b) Is the number of elements in the union of two sets necessarily equal to the sum of the numbers of elements in each set? Justify your response.

(c) Try to obtain a formula for the number of elements in the union of two arbitrary sets. Explain.

8. Prove the following parts of Theorem 1.3-2 by using the method of proof in Example 2.

(a) statement 8
(b) statement 9

1.4 Counting the Elements of a Set

PROBLEM A number of states have adopted a standard form for auto license plates: a six-symbol sequence in which each of the first three symbols can be any letter of the alphabet and each of the last three symbols can be any numerical digit. A witness to a bank robbery can remember only that the license number of the getaway car started with A and ended with 9. How many license numbers could possibly fit this description? ∎

Each of us learns to count at a very early age. This is no small accomplishment, because to begin with we have to learn the *names* of the first several integers and then the *order* in which those names must be used in counting. Later we learn how to write a symbol to represent each number whose name we can say. But few of us ever need to learn the names of all positive integers; for example, what is the name of the integer

$$1205573579033131359447442538767?$$

You might answer, "Who cares?" At the end of this chapter we shall see why that number is very interesting. But the point is that the process of counting is not always simple, even though we may be counting only finite sets. (A substantial portion of this book is devoted to *combinatorics* —the study of ways to count finite sets.)

The key to any counting process is a natural extension of the method we first used. When we want to buy 17 apples, we select one apple at a time while speaking the name of an integer starting with 1, proceeding in sequence until we reach the name "seventeen." In that way we establish what is called a one-to-one correspondence between the selected set of 17 apples and the set of integers $\{1, 2, 3, \ldots, 17\}$, the first 17 positive integers.

A *one-to-one correspondence* between set A and set B is a pairing of the elements of A with the elements of B such that each element of A is paired with one and only one element of B and each element of B is paired with one and only one element of A.

This definition was first given by Cantor, who then used it to distinguish between finite sets and infinite sets by defining an *infinite* set as a set that can be put into one-to-one correspondence with a *proper* subset of itself. For example, a one-to-one correspondence between the infinite set P of all positive integers and the infinite set E of all even positive integers is given by the pairing that relates k with $2k$ for each k in P.

$$P: 1, 2, 3, 4, \ldots$$
$$\updownarrow \quad \updownarrow \quad \updownarrow \quad \updownarrow$$
$$E: 2, 4, 6, 8, \ldots$$

Any nonvoid finite set B has the property that its set of elements can be paired in one-to-one correspondence with the set $\{1, 2, 3, \ldots, m\}$ of all positive integers in sequence from 1 to m, inclusive, for some positive integer m. If we use the symbol $n(B)$ to denote the number of *distinct* elements of B (or the *size* of B), then

$$n(B) = m,$$

where m is described in the previous sentence.

Observe carefully that if

$$B = \{a, a, b, a, c, b, c, a\},$$

then $n(B) = 3$ (rather than 8) (we assume that each of the symbols a, b, and c represents one and only one element).

In this section we shall begin learning how to count more efficiently, starting with the following example.

EXAMPLE   At Joe's Diner the luncheon special includes a sandwich and a choice of soup or salad (but not both). There are five kinds of sandwiches, three soups, and four salads.

1. A customer wants a turkey sandwich, but he is having difficulty deciding whether to choose soup or salad; he is also undecided about which soup or salad he wants.
2. A second customer is undecided about the sandwich, but she can rule out the jellied salad (she tried it on an earlier occasion).

How many choices does each customer have?   ■

The techniques for answering these two questions are different, depending critically on the little words "or" and "and" that Joe used in describing the luncheon special. The first customer may select only one item to eat with the turkey sandwich, and there are seven options available (three soups, four salads). Hence the first customer has seven choices: $7 = 3 + 4$.

This example of the choice of either a soup or a salad leads to the first of two basic rules of counting. If we regard the list of three soups as defining a set A and the list of four salads as defining a set B, then the menu allows one choice from the set $A \cup B$. Since A and B are disjoint sets, $A \cup B$ contains $3 + 4$ elements, which is the way we counted the first customer's choices.

Additive Rule of Or   If A, B are finite sets that are *disjoint* (have no members in common), then the number of different choices of a *single* element from one set *or* from the other set is

$$n(A) + n(B).$$

REMARKS   1. The crucial word here is "or." To choose one element from A or one from B, we must choose one from $A \cup B$. At Joe's Diner the number of choices from the first customer was determined by adding, because of the use of the words "either . . . or" in the menu rule. Note carefully that this use of "or" means "one or the other but not both," called the *exclusive or.* Mathematics typically uses the *inclusive or,* meaning "one or the other or both," unless a specific statement is made to the contrary.

2. The Additive Rule of Or is sometimes called the *Rule of Sum.* We prefer the former name because it tells us both when and how to use this rule, whereas the latter tells us *how* to use this rule but not *when* to use it.

3. In Section 1.5 we consider the same problem for the case in which A and B have a nonempty overlap.   ■

The second customer, however, may choose any sandwich and one of three soups or one of three acceptable salads. There are five choices of sandwiches and six choices of soups or salads. If she chooses a ham sandwich, she has six choices remaining for soup or salad. If she chooses a cheese sandwich, she has the same six options for soup or salad, and she has the same six options whether she chooses a turkey or peanut butter sandwich or a hamburger. Hence the second customer has 30 choices: $30 = 5(6)$.

The second customer's reasoning leads us to a second counting rule. First observe that the second customer needs to make *two* choices: one from the list of sandwiches, and the other from the combined lists of soups and salads. Thus we need to count the number of *pairs* of elements, the first entry of the pair being one sandwich and the second being one soup or one salad. There are five choices for the first entry of the pair and six choices for the second entry, resulting in $30 = 5(6)$ different pairs.

In this example there is no reason to order a sandwich first and a soup or a salad second. The number of choices the second customer has can be regarded as $6(5)$ or $5(6)$. Since the result is the same, it is only for convenience that we arbitrarily choose a fixed order for the categories of menu items.

Before stating the second basic rule of counting, we introduce another way of combining two sets to create another set.

The *cartesian product* $A \times B$ of two sets A and B is the set of all *ordered pairs* (a, b) such that $a \in A$ and $b \in B$. In symbols,

$$A \times B = \{(a, b) \mid a \in A \text{ and } b \in B\}.$$

Two ordered pairs (r, s) and (u, v) are defined to be *equal* if and only if $r = u$ and $s = v$.

The cartesian product $A_1 \times A_2 \times \cdots \times A_m$ of m sets $A_1, A_2, \ldots, A_m$ is the set of all *ordered m-tuples* $(a_1, a_2, \ldots, a_m)$, where $a_i \in A_i$ for $i = 1, 2, \ldots, m$.

A Counting Rule Let A and B be finite sets having $n(A)$ and $n(B)$ elements, respectively. Then $n(A \times B) = n(A)n(B)$.

To convince ourselves of this principle of counting, suppose that $A = \{a_1, a_2, \ldots, a_k\}$ and $B = \{b_1, b_2, \ldots, b_m\}$. Write the elements of $A \times B$ in the following rectangular array:

$$(a_1, b_1), (a_1, b_2), \ldots, (a_1, b_m)$$
$$(a_2, b_1), (a_2, b_2), \ldots, (a_2, b_m)$$
$$\cdot$$
$$\cdot$$
$$\cdot$$
$$(a_k, b_1), (a_k, b_2), \ldots, (a_k, b_m).$$

Each of the k rows of the array lists m elements of $A \times B$ and all elements are in the array, so

$$n(A \times B) = km.$$

Observe that the cartesian product of two sets is a generalized form of the rectangular coordinate system introduced by Descartes for representing each point on a plane by a uniquely determined ordered pair of real numbers (x, y), such that, conversely, each ordered pair of real numbers represents only one point. In this way a cartesian coordinate system establishes a one-to-one correspondence between the points of the plane and the set of all ordered pairs of real numbers.

We will use a more general form of this counting principle in problems that ask us to calculate the number of different outcomes of a process that is carried out in several stages, when the outcome of any stage might affect the possible outcomes of a later stage. This would be the case, for example, if the menu at Joe's Diner indicated that each choice of sandwich was accompanied by a choice of soup or salad from a particular sublist of the seven soup and salad items.

Multiplicative Rule of And Let A and B be finite sets with $n(A) = k$ and $n(B) = m$. Assume that there is a natural number $p \leq m$ such that each element a_i of A determines a subset B_i of B such that $n(B_i) = p$. The number of different choices of ordered pairs (a_i, b_{ij}), where a_i is chosen first from A and then b_{ij} is chosen from the p-element subset B_i of B determined by a_i, is

$$n(A)n(B_i) = kp.$$

It will be useful to record these further observations.

1. When $B_i = B$ for each i, this rule specializes to the rule for the number of elements in the cartesian product $A \times B$.
2. The Multiplicative Rule of And applies without change (except in notation) when $B = A$.
3. The crucial word to look for is "and." The first entry (a_i) is chosen arbitrarily from A, and once that choice is made the second element is chosen from the subset B_i determined by the first element. In general, B_i depends upon the first choice.

To convince yourself concerning this principle of counting, sometimes called the *Rule of Product,* imagine how you could write a list of all possible choices of president and secretary for a club of three women and seven men if tradition requires that the president must be a woman and the secretary a man. Each of the three women could serve with any of the seven men, so the list would contain $3(7) = 21$ different pairs of names.

If the selection of two club officers were not restricted by tradition, the number of choices would be different, but the Multiplicative Rule of And would still apply. Think of electing one officer and then electing the second officer from among the members not chosen for the first office. There are 10 possible outcomes of the first election, but only 9 for the second. Hence, the list of different election outcomes contains $10(9) = 90$ pairs of names. In this example the sets A, B, and B_i are chosen as follows: Each set A and B is the set of club members, and B_i is the set of all club members except member a_i. The set B_i is chosen in this way because if a_i is elected as president, then a_i is ineligible for secretary. Of course, if one person may hold both offices at the same time, then there are $10(10) = 100$ possible outcomes.

The Additive Rule of Or can be extended to any finite number of finite sets.

Let A_1, A_2, . . . , A_m be finite sets. If each pair of sets is disjoint ($A_i \cap A_j = \varnothing$ for each $i \neq j$), the number of different choices of one element from the set $A_1 \cup A_2 \cup \cdots \cup A_m$ is

$$n(A_1) + n(A_2) + \cdots + n(A_m).$$

The Multiplicative Rule of And can also be extended to any finite number of finite sets. Because it is rather complicated, we state here only the following special case.

Let $A_1, A_2, \ldots, A_m$ be finite sets. Then the number of different choices of ordered m-tuples $(a_1, a_2, \ldots, a_m)$, where $a_i \in A_i$ for $i = 1$, $2, \ldots, m$, is

$$n(A_1)n(A_2) \cdots n(A_m).$$

Now you should be able to solve the license plate problem posed earlier in this section. A license plate is a 6-tuple (three letters followed by three digits). The problem asks you to count how many ways you can fill in the blanks in the figure given that the second and third squares may contain any letter, and the fourth and fifth squares may contain any digit. See Exercise 4.

A					9

Exercises 1.4

1. A pet store has 20 dogs, 30 cats, and 9 turtles.
   (a) In how many ways can you choose one pet if that pet must be a dog or a cat?
   (b) In how many ways can you choose two pets if the first must be a dog and the second a cat?
   (c) In how many ways can you choose two pets if the first must be a dog and the other must be a cat or a turtle?
   (d) In how many ways can you choose three pets: first a dog, then a cat, and then a turtle?

2. Let $A = \{1, 2, 3\}$, $B = \{3, 4\}$. List the members of the following sets.
   (a) $A \times B$
   (b) $B \times A$
   (c) $B \times B$
   (d) $A \times (A \cap B)$

3. Let A, B, C be three sets, no two of which have an element in common. Let $n(A) = k$, $n(B) = m$, and $n(C) = p$.
   (a) In how many ways can you choose an ordered pair with the first entry from $A \cup B$ and the second from C?

(b) In how many ways can you choose a single element from either the set A or the set $B \times C$? What further information is needed about the sets A and $B \times C$ to answer this question?

(c) How many elements are in each of the following sets: $A \cup A$, $B \times B$, $(A \cup C) \times (B \cup C)$, $A \times B \times C$?

4. Solve the license plate problem stated in the first paragraph of this section.

5. Solve Exercise 4, assuming that the witness also remembers that no letter of the license number was repeated.

6. For this exercise you may refer to the definitions of prime integer and composite integer given in Section 1.2.

(a) In how many ways can you choose an ordered pair of distinct prime numbers, given that each chosen prime number must be less than 30?

(b) In how many ways can you choose an ordered 3-tuple of distinct prime numbers, given that each chosen prime number must be less than 30?

(c) In how many ways can you choose an ordered pair of elements, with the first entry being an ordered pair of the kind described in **a** and the second entry being an ordered 3-tuple of the kind described in **b**?

(d) If

$$P = \{y \in N \,|\, y \text{ is prime and } y < 30\},$$
$$C = \{w \in N \,|\, w \text{ is composite and } w < 30\},$$

determine the number of elements in the set $(P \times P) \times (C \times C \times C)$.

7. Answer the question posed in the last line of the following nursery rhyme.

As I was going to St. Ives,
I met a man with seven wives,
Each wife had seven sacks,
Each sack had seven cats,
Each cat had seven kits:
Kits, cats, sacks, wives,
How many were going from St. Ives?

(The original rhyme asks a different question: How many were going *to* St. Ives? What is the answer to the original question?)

8. A string of four letters is to be formed. If there are 10 different letters to choose from and no letter may be used more than once, how many different strings may be formed? (An example of a string of four letters is *e a m h;* it is considered different from the string *e a h m.*)

9. (a) List all the subsets of a set containing three elements.
   (b) List all the subsets of a set containing four elements.
   (c) To count the number of subsets of a set $W = \{a_1, a_2, \ldots, a_n\}$ containing n elements we can proceed as follows. With each subset of W we associate a decision table.

a_1	a_2	a_3	$\cdots$	a_n
			$\cdots$	

Each of the lower boxes will contain either the symbol 0 or the symbol 1. The symbol 0 below a_i will mean that a_i is not in the subset, whereas the symbol 1 below a_i will mean that a_i is included in the subset. This implies that the number of different choices of ordered n-tuples, having 0 or 1 as each entry, is equal to the number of subsets of the set W.

   Count the number of different choices of ordered n-tuples having 0 or 1 as each entry and thus obtain a formula for the number of subsets of a set containing n elements.

10. How many four-digit integers are there with no digit repeated?

11. How many four-digit integers do not have 9 as a final digit?

1.5   The Inclusion-Exclusion Principle

PROBLEM   In a survey concerning musical preference, respondents were asked to check their favorites from a list of 50 classical composers. Exactly 100 respondents checked at least one of the three composers, J. S. Bach, Aaron Copland, and Claude Debussy. Of those 100 respondents

   70 checked Bach
   60 checked Copland
   25 checked Debussy

35 checked Bach and Copland

15 checked Copland and Debussy

2 checked Bach, Copland, and Debussy

How many respondents checked Bach and Debussy but not Copland?

■

Our objective in this section is to introduce and to illustrate additional counting techniques that will enable you to answer this question (see Exercise 3) and to solve similar counting problems.

In Example 1 (and elsewhere throughout this book) we refer to the concept of divisibility in the set of integers, which is defined as follows:

If a and b are integers and if $a \neq 0$, then we say that b *is divisible by a* if and only if there is an integer x such that $b = ax$. Other ways to express that b is divisible by a are to say that b *is a multiple of a*, or that a *divides b*.

Example 1 demonstrates that to count the number of elements in a subset A of a set S, it is sometimes easier to count the complement A' of A in S and then to use the equation $n(A) = n(S) - n(A')$.

EXAMPLE 1   How many integers from 1 to 945, inclusive, are not divisible by 3?
Let $S = \{x \in I \mid 1 \leq x \leq 945\}$ and $A = \{x \in S \mid x \text{ is not divisible by 3}\}$. Then A' is the set of numbers in S that are divisible by 3. Because exactly every third number of S is divisible by 3 and because $945/3 = 315$, we have

$$n(A') = 315 \quad \text{and} \quad n(A) = n(S) - n(A') = 945 - 315 = 630.$$

Thus 630 integers from 1 to 945, inclusive, are not divisible by 3.   ■

One hazard in counting the elements of a set is duplication — counting an element more than once. In counting small finite sets of physical objects, we can avoid duplication by marking each object as it is counted. But that wouldn't be practical for large sets. Thus sometimes it is easier not to *avoid* duplication but to keep track of the duplications and *compensate* accordingly. Examples 2 and 3 show how this can be done.

EXAMPLE 2   How many integers from 1 to 945, inclusive, are divisble by either 3 or 5?

Because of the word "or," we are inclined to use the Additive Rule of Or to answer this question. To do so, we first determine how many of the given integers are divisible by 3 and how many are divisible by 5. From Example 1 we know there are 315 integers divisible by 3; because every fifth integer is divisible by 5 and because 945/5 = 189, exactly 189 of the given integers are divisible by 5. If we were to apply the Additive Rule, we would obtain 315 + 189 = 504 as the answer. But we are not justified in applying the Additive Rule because the set of integers divisible by 3 and the set of integers divisible by 5 are *not disjoint*. The numbers 15, 30, 45, . . . , 945 belong to both of those sets, so we have counted each of those numbers exactly twice. To compensate for that duplication, we need to subtract the number of integers between 1 and 945 that are divisible by 15. Because exactly every fifteenth integer is divisible by 15 and because 945/15 = 63, we subtract 63 from the previous total, thereby obtaining

$$315 + 189 - 63 = 441.$$

Thus the correct answer to the question is that 441 integers between 1 and 945, inclusive, are divisible by either 3 or 5.   ■

EXAMPLE 3   How many integers between 1 and 945, inclusive, are divisible by either 3 or 5 or 7?

We begin, as in Example 2, by overcounting and then compensating. Exactly 315 of the given integers are divisible by 3, 189 are divisible by 5, and 945/7 = 135 are divisible by 7. Thus our initial count is 639, the sum of 315, 189, and 135. Next we begin compensating by subtracting the numbers of integers divisible by both 3 and 5, by both 3 and 7, and by both 5 and 7. Noting that 945/15 = 63, 945/21 = 45, and 945/35 = 27, we obtain

$$(315 + 189 + 135) - (63 + 45 + 27) = 639 - 135 = 504$$

as a second approximation of the desired count. However, further compensation is necessary, because each number divisible by 105 (the product of 3, 5, and 7) was counted three times in the initial count of 639 and then was counted three times in the compensating count of 135. Exactly 945/105 = 9 such numbers were removed by overcompensation in the second approximation, 504, of the desired count. Hence we must count these numbers once more, and we obtain a final count of

$$(315 + 189 + 135) - (63 + 45 + 27) + 9 = 513.$$

We conclude that there are 513 integers between 1 and 945, inclusive, that are divisible by either 3, 5, or 7.   ■

We now formalize the principles used in Examples 2 and 3. Let S denote a set, and let A, B, and C be finite subsets of S. The principle used in Example 2 can be expressed by the formula

$$n(A \cup B) = n(A) + n(B) - n(A \cap B),$$

where A represents the set of given numbers divisible by 3, and B represents those divisible by 5. Here a Venn diagram directs our thinking very effectively. See Figure 1.5-1.

There are $n(A)$ elements in set A and $n(B)$ elements in set B. Each of these counts includes the elements of $A \cap B$, so $n(A) + n(B)$ counts each element of $A \cup B$ at least once but counts each element of $A \cap B$ twice (once too often). Because this counting formula is important, we state it as a theorem.

Theorem 1.5-1 If A and B are finite subsets of a set S, then

$$n(A \cup B) = n(A) + n(B) - n(A \cap B).$$

Before formalizing the principle used in Example 3, we recall that set union and set intersection are associative. Thus we can write both $A \cup (B \cup C)$ and $(A \cup B) \cup C$ without parentheses as $A \cup B \cup C$. Similarly, we can write $A \cap (B \cap C)$ and $(A \cap B) \cap C$ as $A \cap B \cap C$. Then the reasoning used in Example 3 can be expressed by the following formula.

$$n(A \cup B \cup C) = n(A) + n(B) + n(C)$$
$$- [n(A \cap B) + n(A \cap C) + n(B \cap C)] + n(A \cap B \cap C),$$

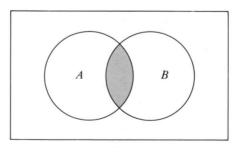

Figure 1.5-1

where A represents the set of given integers divisible by 3, B those divisible by 5, and C those divisible by 7. To convince yourself of this formula, refer to the Venn diagram in Figure 1.5-2. The sum $n(A) + n(B) + n(C)$ counts each element of $A \cup B \cup C$ at least once, counts each element of regions 1, 2, and 3 twice, and counts each element of region 4 thrice. The altered count

$$n(A) + n(B) + n(C) - [n(A \cap B) + n(A \cap C) + n(B \cap C)]$$

no longer counts each element of $A \cup B \cup C$ at least once. It does count each element in $A \cup B \cup C$, except those in region 4, exactly once, but it fails to count any elements in region 4 because each element in region 4 was counted three times in $n(A) + n(B) + n(C)$ and then subtracted three times by the terms enclosed in brackets in the altered count. Thus if we now add $n(A \cap B \cap C)$ to the altered count, we get a correct formula for $n(A \cup B \cup C)$.

This formula for $n(A \cup B \cup C)$ also can be deduced algebraically, using the algebra of sets and the formula for $n(A \cup B)$. You are asked to do this in Exercise 6.

The two formulas for counting the number of elements in the union of two or three finite sets are specific instances of the general *inclusion-exclusion principle*.

General Inclusion-Exclusion Principle Let n be any positive natural number. The number of elements in the union of any n finite sets can be obtained by the following procedure:

1. Calculate the sum of the numbers of elements in all n individual sets to obtain a temporary total T.

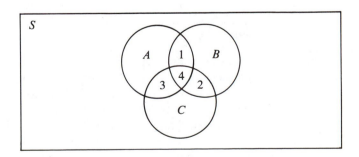

Figure 1.5-2

2. Calculate the sum of the numbers of elements in the intersection of each pair of the sets. Subtract that sum from the current value of T to obtain a new current value of T.

3. Calculate the sum of the numbers of elements in the intersection of each triple of the sets. Add that sum to the current value of T to obtain a new current value of T.

4. Calculate the sum of the numbers of elements in the intersection of each quadruple of the sets. Subtract that sum from the current value of T to obtain a new current value of T.

.

.

.

Continue in this manner, alternately adding and subtracting from the current value of T, until at step n the final correction will be the number of elements in the intersection of all n of the sets. That number is to be added to the previous value of T if n is odd but is to be subtracted if n is even.

EXAMPLE 4   A group of 28 students plans to study languages in Switzerland for the year. Nineteen of the students speak French, and 13 students speak German. How many students speak both German and French, given that a condition for enrollment in the program is that each applicant must speak either French or German?

Let S denote the set of 28 students, F the French-speaking subset of S, and G the German-speaking subset. Then $S = F \cup G$, because each applicant must speak either French or German, and

$$28 = n(S) = n(F \cup G) = n(F) + n(G) - n(F \cap G)$$
$$= 19 + 13 - n(F \cap G).$$

Hence, $n(F \cap G) = 19 + 13 - 28 = 4$, so four students in the program can speak both French and German.   ■

EXAMPLE 5   Conducting a study of employment patterns by interviewing from door to door, a sociology student obtained information about 972 persons from 360 households. Her report stated that 630 were older than 16 years of age, of these 498 were employed, and 197 were employed females. There were 62 unemployed females over 16. There were 172 females not older than 16, and no person aged 16 or younger was employed. From this information is it possible to determine the number of persons in each of the categories related to this survey?

The initial difficulty here is the confusing way in which the data is presented. To clarify matters, let

$$S = \text{set of persons surveyed,}$$
$$A = \{x \in S \mid x \text{ is older than 16 years of age}\},$$
$$F = \{y \in S \mid y \text{ is female}\},$$
$$E = \{z \in S \mid z \text{ is employed}\}.$$

Observe that the complement F' is the set of males in the survey, and an interpretation of both A' and E' is equally clear. There are three attributes of classification (age, sex, and employment status) and two categories within each attribute, so we can present the information in the form of a table by choosing one attribute (sex, for example) as label for the columns, a second attribute (employment) to label the rows, and the third attribute (age) to label subcolumns. Because some data is given for combined categories, we also include a column for females and males combined.

Sex	Female F			Male F'			Combined $F \cup F'$		
Age Employment	A	A'	Sum	A	A'	Sum	A	A'	Sum
E		*	197	*	*			0	498
E'	62								*
Sum		172					630	*	972

Table 1.5-1

You should read the problem again, verifying that each piece of data is entered correctly. (We regard the number of households visited to be irrelevant to our objective.) Now the problem is easy to complete; the five positions in Table 1.5-1 that are marked with an asterisk can be filled in first, yielding $n(E') = 474$, $n(A') = 342$, $n(F' \cap E) = 301$, $n(F \cap A' \cap E) = 0$, and $n(F' \cap A' \cap E) = 0$. The first two entries allow us to complete the combined columns, and the fourth allows us to complete the female columns; the male columns are then easy to obtain. ∎

Exercises 1.5

1. Among a group of 1000 college students, 40 own a car and 30 own a home computer. Sixty-two own either a car or a home computer or both.

   (a) How many own both a car and a home computer?
   (b) How many own a car but not a home computer?

2. The 90 rooms at a motel are classified in three ways, smoking or nonsmoking, with or without a TV set, and two-bed or one-bed. There are 48 smoking rooms, 75 TV rooms, and 70 two-bed rooms. Forty of the smoking rooms have TV sets, and 32 of the two-bed rooms permit smoking. Sixty of the two-bed rooms have a TV set. Each room except one either has a TV set, or has two beds, or permits smoking.

   (a) How many of the two-bed rooms have a TV set and permit smoking?

   (b) How many of the one-bed rooms permit smoking but have no TV set?

3. Solve the problem concerning musical preference stated at the beginning of this section.

4. A dealer received a shipment of 40 ovens. The options available for this kind of oven are a black glass door, a self-cleaning feature, and a built-in rotisserie. In this shipment 10 of the ovens have black doors, 24 are self-cleaning, and 6 have rotisseries. Four of the 40 ovens have all three options. Show that the maximum number of ovens that have one or more options is 32. (Note that $A \cap B \cap C \subseteq A \cap B$ for any three sets A, B, C.)

5. Four kinds of medicine are frequently given to patients at a certain hospital. We denote the four kinds of medicine by the letters A, B, C, and D. The hospital provides the following data.

Kinds of Medicine	Number of Patients Taking the Medicine
A	160
B	65
C	35
D	50

Some patients take more than one of the four medicines. According to records, the only overlaps are given in the following list.

A and B	55
A and C	30
A and D	50
B and C	6
A, B, and C	2

(a) Let $P(A)$, $P(B)$, $P(C)$, $P(D)$ denote the sets of patients taking medicines A, B, C, and D, respectively. Draw a Venn diagram showing the sets $P(A)$, $P(B)$, $P(C)$, $P(D)$ and the number of elements in each subregion.

(b) Assuming that every patient takes at least one of the four kinds of medicine, use the inclusion-exclusion principle to determine the number of patients in the hospital.

6. Derive the inclusion-exclusion principle for three sets from the corresponding formula for two sets, as given in Theorem 1.5-1.

7. Complete Table 1.5-1.

8. A political science class has a total enrollment of 59 students of whom 28 are seniors. Thirteen of the nonsenior men received a course grade of B or above, and four of the nonsenior women received a grade below B. Forty students received a grade of B or above; of these 40 students, 18 were seniors. Fifteen of the seniors enrolled in the class were women, and of these 15 students 12 received a grade of B or above. Construct a table similar to Table 1.5-1.

9. Apply the general inclusion-exclusion principle to write a formula for the number of elements $n(A \cup B \cup C \cup D)$ in the union of four sets.

10. How many integers from 1 to 3080, inclusive, are divisible by 5 or 7 or 11?

11. How many integers from 1 to 1050, inclusive, are divisible by neither 3 nor 5?

12. Let S denote the set of all four-digit natural numbers that can be formed using the digits 1, 2, 3, . . . , 9, allowing repetition of digits.

(a) How many numbers in S have at least one digit that is 2?

(b) How many numbers in S have at least one digit that is 2 or at least one digit that is 3?

METHODS OF REASONING

1.6 An Algebra of Statements

To develop an algebra of sentences for use in deductive reasoning, we need to restrict our consideration to a special kind of sentence called a *statement* or *proposition.* We must exclude from consideration any declaration that cannot be assigned exactly one of the two labels T and F. For example, consider the sentence

S: This sentence is false.

If the sentence S is declared to be true, then it is true that S is false, a contradiction. But if S is declared to be false, then it is false that "this sentence is false," so S must be true, another contradiction. Moreover, nonsense sentences, such as this declaration of the Jabberwocky, cannot sensibly be labeled either T or F:

All mimsy were the borogoves,
and the mome raths outgrabe.

Neither of these two examples of sentences satisfies the following definition of a statement.

A *statement* (or *proposition*) is a declarative sentence that is either true or false but not both.

Observe that this definition does not require us to judge the truth or falsity of a sentence but only to verify that the sentence must be one or the other and not both. Indeed, we do not attempt to define "true," nor is it necessary for us to do so.

From your geometry course in school you will recall that Euclid and his followers provided the world with its first example of a deductive system by starting from a foundation of statements that were called *axioms* and building a large body of statements called *theorems* by applying specified rules of deduction. To the Greeks an axiom was a "self-evident truth," and conclusions that were deduced from the axioms were facts of geometry. But now it is known that there are geometries different from Euclid's with implications that are not Euclidean facts. Therefore most contemporary mathematicians regard axioms as assumptions that are true statements for the study at hand.

The end of Section 1.3 called particular attention to the overriding importance of the little words "or," "and," and "not" in the definition of set operations. "Or" and "and" are conjunctions that can be used to join two statements to form a third statement, whereas "not" can be used to convert any statement into a different statement. For example, from the statements

1. Today is Saturday.
2. Sam is playing ball.

we can form the additional statements

    3. Today is Saturday or Sam is playing ball.

    4. Today is Saturday and Sam is playing ball.

    5. Today is not Saturday.

In Section 1.3 the word "or" was used to define set union: the set $A \cup B$ consists of all elements that belong either to A or to B or to both. By analogy, when the word "or" is used to combine two given statements into a third statement, the combined statement is regarded as true if and only if *at least one* of the given statements is true. Thus in the algebra of statements, as in the algebra of sets, the word "or" is used to convey the idea of "at least one."

The word "and" is used in set theory to convey the idea of "both," because the set $A \cap B$ consists of all elements that belong both to A and to B. In the algebra of statements the word "and" is used to combine two given statements into a third statement, and the third statement is regarded as true if and only if *both* of the given statements are true.

The word "not" is used in set theory to define the complement of a set A relative to a second set S: the complement A' of A in S consists of all elements of S that are not in A. In the algebra of statements the word "not" is used to convert a given statement into its negation. The negation of a given statement is regarded as true if and only if the given statement is false.

EXAMPLE 1   Considering the foregoing list of numbered statements, we note that statement 3 will be true on any Saturday; it also will be true at any time Sam is playing ball. It is false only when Sam is not playing ball on a day other than Saturday. Statement 4 is true only when Sam is playing ball on Saturday. Statement 5 is true on every day of the week except Saturday. ∎

This informal discussion has attempted to focus attention on three little words ("and," "or," "not") which are fundamentally important in set theory, and which derive their meanings from the language that we use for verbal communication. Indeed, those same words can be used to develop an algebra of statements that precisely duplicates the algebra of sets. An algebra of statements is introduced not so much to display an interesting parallel between set theory and formal logic as to demonstrate means of analyzing the precise meaning of a compound statement in terms of the meanings of its component parts. Moreover, we need to know how to transform a convoluted statement into a simpler one having the same meaning, just as we use the algebra of numbers to transform and to simplify complex algebraic expressions.

The analogy to the algebra of sets will become more evident if we represent each given statement by a single letter. For example, let p and q each represent a statement. In place of the word "or" we use the connective symbol $\vee$ (analogous to $\cup$), in place of "and" we use the symbol $\wedge$ (analogous to $\cap$), and in place of "not" we use the symbol $\sim$, which can also be read as "it is not the case that." In that notation statements 3, 4, and 5 can be written in symbolic form as

3. $p \vee q$ (p or q),
4. $p \wedge q$ (p and q),
5. $\sim p$ (it is not the case that p),

where p denotes statement 1 and q denotes statement 2.

Mathematics is especially concerned with *implications* — statements of the form

if h then c (also phrased as "h implies c"),

where h and c are statements. To claim that this statement is true means that

c is true whenever h is true.

That fails to be the case only when h is true and c is false. We therefore regard "if h then c" to be true if and only if the following statement is true:

Either c is true or h is false.

This is all that "h implies c" means. It does not mean that h is true nor that h is false, nor does it mean that c is true nor that c is false.

An implication that has been proved to be true is called a *theorem*.

The convention of regarding "h implies c" to be true whenever h is false is sometimes puzzling, but the following example of implication in everyday use should clarify this convention.

If the temperature is subfreezing, then I will stay home and read by the fire.

We certainly would not want to label this statement as false when the temperature is above freezing, a condition about which the statement is silent. We shall give another reason for this convention in Section 1.7.

What is important to realize is that in a given study each statement p

can have either of two values. Those values can be indicated by any of the symbol pairs

true, false

T, F

1, 0

valid, invalid

orange, green.

Arbitrarily we choose to use 1 and 0 as the only possible values of a statement. Each axiom (a) is given the value 1, and its negation ($\sim a$) is assigned the value 0. More generally, if p is a statement whose value is not known, then the value of $\sim p$ is defined to be the opposite of the value of p. Table 1.6-1 gives this concisely.

For a statement constructed from a pair of given statements (p, q), there are two possible values, 1 and 0, for p and independently the same two possible values for q, making four possible pairs of values for the pair. Thus the values of $p \vee q$ and $p \wedge q$ are defined by Table 1.6-2. Observe that "p or q" has the value 1 if and only if 1 is the value of p, or of q, or of both. But "p and q" has the value 1 if and only if 1 is the value of both p and q. "Not p" has the value p does not have.

From these tables we can compute a table of values for any statement involving a finite number of sentences, joined in an appropriate manner by the symbols $\vee$, $\wedge$, and $\sim$. "Appropriate manner" means two things: (1) the string of symbols must be a statement, and (2) parentheses must be used where necessary to make the statement unambiguous. For example, $q \sim p$ is not a statement because $\sim$ can be used to negate the next statement but not to connect two statements. The sentence

$$r \vee q \wedge \sim p$$

is a statement but is ambiguous because parentheses are needed to indicate the sequence of operations. Does it mean

$$(r \vee q) \wedge \sim p \quad \text{or} \quad r \vee (q \wedge \sim p)?$$

p	$\sim p$
1	0
0	1

Table 1.6-1

		p or *q*	*p* and *q*
p	*q*	$(p \lor q)$	$(p \land q)$
1	1	1	1
1	0	1	0
0	1	1	0
0	0	0	0

Table 1.6-2

To see that these two sentences have different meanings (different value tables), we calculate the value table of each. Columns 6 and 8 of Table 1.6-3 are not identical, so the two associated sentences have different meanings for certain choices of *p*, *q*, and *r*. To be specific, let *p*, *q*, and *r*, respectively, be the statements: 2 is an even integer, 3 is an odd integer, 5 is an odd integer. Then the statement $(r \lor q) \land \sim p$ is false, whereas $r \lor (q \land \sim p)$ is true. That is, the two statements have different values for this choice of *p*, *q*, and *r*.

Some pairs of statements will have the same value tables even though the two statements appear different. It is convenient to have terminology and notation for such pairs, as defined next.

Two statements are said to be *logically equivalent* if and only if the two statements have the same value for every possible assignment of values to their components. To denote that statements *s* and *t* are logically equivalent, we write

$$s \equiv t.$$

p	*q*	*r*	$(r \lor q)$	$\sim p$	$(r \lor q) \land \sim p$	$q \land \sim p$	$r \lor (q \land \sim p)$
1	1	1	1	0	0	0	1
1	1	0	1	0	0	0	0
1	0	1	1	0	0	0	1
1	0	0	0	0	0	0	0
0	1	1	1	1	1	1	1
0	1	0	1	1	1	1	1
0	0	1	1	1	1	0	1
0	0	0	0	1	0	0	0

Table 1.6-3

It will not surprise you to learn that the statement operators $\vee$ and $\wedge$ are both commutative and associative: For all statements p, q we have

$$(p \vee q) \equiv (q \vee p) \qquad \text{and} \qquad (p \wedge q) \equiv (q \wedge p),$$
$$p \vee (q \vee r) \equiv (p \vee q) \vee r \quad \text{and} \quad p \wedge (q \wedge r) \equiv (p \wedge q) \wedge r.$$

This is due to the meanings of "or" and "and" and is easily verified by a value table. Also, each is distributive over the other:

$$p \vee (q \wedge r) \equiv (p \vee q) \wedge (p \vee r),$$
$$p \wedge (q \vee r) \equiv (p \wedge q) \vee (p \wedge r).$$

You may verify one or both of these as exercises, and similarly you may verify the pair

$$p \vee p \equiv p \quad \text{and} \quad p \wedge p \equiv p.$$

Undoubtedly you have observed something familiar about these assertions. If you refer to Section 1.3, you will find that they are thinly disguised copies of statements 7 through 10 and 7* through 10* of Theorem 1.3-2. Because the other two parts of that theorem (11 and 11*) establish the existence of identity elements for the algebra of sets, it is reasonable to look for a statement that might be an identity element relative to $\vee$ in the algebra of statements. We seek a statement z such that the following holds for every statement p

$$p \vee z \equiv p.$$

The requirement that z must satisfy this relation for *every* p tells us that z must be universal and hence defined independently of any particular p. Thus we see that z can be any statement whose value table has 0 for every entry; an example of such a z is

$$q \wedge (\sim q)$$

where q is any statement. Any statement z of this form is called a *contradiction,* and all contradictions are logically equivalent. Its role in logic is analogous to the role of the empty set in the algebra of sets, because

$$p \vee z \equiv p \qquad \text{for every } p.$$

In contrast, the statement

$$q \vee (\sim q)$$

has a value table with 1 for every entry. Such a statement u is called a *tautology,* and all tautologies are logically equivalent. Note that u acts as an identity element relative to $\wedge$ in the algebra of statements;

$$p \wedge u \equiv p$$

for every statement p. Thus a tautology u in the algebra of statements is analogous to S in the algebra of subsets of S. The operation $\sim$ (negation) appears to be the counterpart in logic to complementation in set theory because for all p

$$p \vee (\sim p) \equiv u,$$
$$p \wedge (\sim p) \equiv z,$$

which are analogous to statements 14 and 14* in Theorem 1.3-3. The letters z and u are chosen to remind us of the arithmetic counterparts zero (the identity of addition) and unity (the identity of multiplication). Also

$$\sim(\sim p) \equiv p,$$

which is analogous to statement 12 of the same theorem.

The De Morgan laws (15 and 15*) also apply to the algebra of statements:

$$\sim(p \vee q) \equiv (\sim p) \wedge (\sim q),$$
$$\sim(p \wedge q) \equiv (\sim p) \vee (\sim q).$$

EXAMPLE 2   To underscore the method of negating compound statements, let's apply the De Morgan laws to the sentences given at the beginning of this section. The negation (or denial) of statement 3,

> Today is Saturday   or   Sam is playing ball,

is

> Today is not Saturday   and   Sam is not playing ball.

The negation of statement 4,

> Today is Saturday   and   Sam is playing ball,

is

> Today is not Saturday   or   Sam is not playing ball.   ∎

Some of the most frequent errors of logic that occur in printed and spoken language occur because the word "not" is used incorrectly, so you are urged to learn the De Morgan laws and to learn how to use them properly.

Thus far we have established for the algebra of statements all the facts of the algebra of sets that appear in Theorem 1.3-2 and all except number 13 of those in Theorem 1.3-3. Number 13 and each statement of Theorem 1.3-1 involve the subset relation for sets, and we have not yet identified a corresponding concept in logic. Section 1.7 establishes the parallel relationship between these two concepts and examines their application to methods of deductive reasoning.

Exercises 1.6

1. Let p represent the statement, "A is a subset of B," and let q represent "B is a subset of A." Write the following sentences symbolically, using the symbols p, q, $\sim$, $\wedge$, and $\vee$.

    (a) The set A is a subset of B, but B is not a subset of A.
    (b) It is not true that A and B are subsets of each other.
    (c) Neither A nor B is a subset of the other.
    (d) It is not the case that either A is a subset of B or B is a subset of A.
    (e) Either A is a subset of B or B is a subset of A, but not both.
    (f) There is an element of B that is not a member of A.
    (g) The set A is not equal to the set B.

2. Let p represent the statement, "Dr. Jones is at least 35 years old," let q represent "Dr. Jones has published at least five research papers," and let r represent "Dr. Jones is an associate professor." Write the following sentences symbolically, using the symbols p, q, r, $\sim$, $\wedge$, and $\vee$.

    (a) Dr. Jones is not yet 35 years old but has published at least five research papers.
    (b) Dr. Jones is an associate professor or is at least 35 years old, or both.
    (c) It is not the case that Dr. Jones is an associate professor who has published five or more research papers.
    (d) Dr. Jones is either less than 35 years old or an associate professor, but not both.
    (e) Dr. Jones is an associate professor who is not yet 35 years old or who has published fewer than five research papers.

3. Which of the following sentences are statements? Give a brief justification for each answer.

    (a) There are three positive integers a, b, c such that $a^2 + b^2 = c^2$.
    (b) There are three positive integers a, b, c such that $a^2 + b^2 = -c^2$.

    (c) Intersections drink unions.

    (d) No person has set foot on the moon.

    (e) Please do your homework.

    (f) Every positive integer can be expressed as the sum of squares of four integers.

    (g) The sentence I am now writing is false.

    (h) If n is an integer greater than two, then there are no positive integers a, b, c such that $a^n + b^n = c^n$.*

4. Use the De Morgan laws to state in words the negation of each of the following statements. Simplify the negation whenever possible.

    (a) The given integer is a prime or a power of a prime.

    (b) The given integer is divisible by both 2 and 3.

    (c) The given equation has no real root, or it has exactly one real root.

    (d) The given equation has a real root and no more than two imaginary roots.

    (e) The given set is infinite, or it contains fewer than 1000 members.

    (f) The given set of natural numbers contains all natural numbers greater than 1000 and at least one natural number less than or equal to 1000.

5. Calculate value tables for the following statements.

    (a) $q \vee (\sim p)$.

    (b) $[\sim(p \wedge q)] \wedge (p \vee q)$.

    (c) $[\sim(p \vee q)] \wedge p$.

6. Verify the following logical equivalences by the use of value tables.

    (a) $[p \wedge (q \wedge r)] \equiv [(p \wedge q) \wedge r]$.

    (b) $[p \vee (q \wedge r)] \equiv [(p \vee q) \wedge (p \vee r)]$.

    (c) $[\sim(p \wedge q)] \equiv [(\sim p) \vee (\sim q)]$.

    (d) $[p \vee (q \wedge \sim q)] \equiv p$.

    (e) $[p \wedge (p \vee q)] \equiv p$.

7. Let p represent the statement, "The sun is shining"; q the statement, "I read a book"; r the statement, "I play tennis." Translate the following statements into English, simplifying as much as possible.

    (a) $q \wedge \sim p$.

    (b) $\sim(q \vee r)$.

    (c) $[\sim(q \wedge r)] \wedge (q \vee r)$.

---

* This sentence is commonly referred to as *Fermat's Last Theorem* because it was asserted to be true by Pierre-Simon de Fermat, a seventeenth-century French mathematician. Fermat, however, gave no proof, and to this day the assertion has neither been proved true nor has it been proved false.

(d) $[\sim(q \vee \sim r)] \vee (r \wedge \sim q)$.

(e) $\sim[(\sim p \wedge \sim q) \wedge (p \vee r)]$.

8. Let u be a tautology and z a contradiction. Which of the following statements are tautologies, and which are contradictions? If a statement is neither a tautology nor a contradiction, find a simpler statement that is logically equivalent to the given one.

(a) $u \wedge z$.

(b) $u \vee z$.

(c) $(u \wedge p) \vee z$.

(d) $\sim u$.

(e) $\sim z$.

(f) $(p \wedge \sim z) \vee (p \wedge \sim u)$.

(g) $(p \wedge q \wedge r) \vee (\sim p \wedge q \wedge r) \vee \sim q \vee \sim q \vee \sim r$.

1.7   Theorems and Direct Proofs

Before proceeding to develop the algebra of statements that was introduced in Section 1.6, let us summarize our approach to this point by recalling that both the algebra of sets and the algebra of statements have been based upon an intuitive understanding of language — in particular, upon the meanings of the words "or," "and," "not." Although those two algebraic systems use different terminology and different notation, the computational properties of the systems seem to be the same.

This section will extend our understanding of the algebra of statements, particularly with regard to implications, theorems, tautologies, and contradictions, all of which were defined in Section 1.6. We shall then describe and illustrate a type of formal argument frequently used in mathematics to prove theorems. Although formal proofs are not strongly emphasized in this introduction to discrete mathematics, experience in writing proofs is very valuable in learning to think rigorously, to express ideas clearly and precisely, and to understand mathematics.

First recall that an implication is a statement of th form

if h then c,

where both h and c are statements. For example:

1. If x is a real number such that $x^2 = 1$, then $x = 1$ or $x = -1$.

2. If x and y are negative numbers, then xy is a positive number.

Thus the form

if _____ then _____

is a means of combining any two statements to form a third statement.

Also recall the meaning of an implication, as described in Section 1.6 and as summarized in the following definition.

The statement

if p then q

is defined to mean the same as "q is true or p is false"; that is,

q or not p.

"If p then q" also is expressed in the form

p implies q

and is written symbolically as

$$p \Rightarrow q.$$

(The notation $p \rightarrow q$ is sometimes used instead of $p \Rightarrow q$; we shall use only the double-line arrow to denote "implies.")

This definition enables us to write a value table for $p \Rightarrow q$ as shown in the last column of Table 1.7-1. Observe carefully that $p \Rightarrow q$ is false *only when p is true and q is false*. In particular, $p \Rightarrow q$ is true whenever p is false regardless of whether q is true or false. This means that if we reason from a false statement, there can be no guarantee that any subsequent statement is true, even when each step of the argument is logically valid.

We now give an alternative explanation for why the value of $p \Rightarrow q$ should be 1 whenever p has the value 0. It is reasonable to require that the statement $r \Rightarrow (r \vee s)$ be a tautology; that is, it should have the value 1 for every possible assignment of values to r and s. In particular, it should have

p	q	$\sim p$	$q \vee \sim p$	$p \Rightarrow q$
1	1	0	1	1
1	0	0	0	0
0	1	1	1	1
0	0	1	1	1

Table 1.7-1

the value 1 when *r* has the value 0 and *s* has the value 0, and also when *r* has the value 0 and *s* has the value 1. In both cases *r* has the value 0, but $r \lor s$ has the value 0 in the first case and the value 1 in the second case. Thus for $r \Rightarrow (r \lor s)$ to be a tautology, we are required to assign the value 1 to $p \Rightarrow q$ whenever *p* has the value 0.

Observe that the two statements $p \Rightarrow q$ and $q \Rightarrow p$ have different value tables and therefore are different statements. Sometimes people use one of these statements when the other is intended; be sure that you understand the difference. The implication $q \Rightarrow p$ is called the *converse* of $p \Rightarrow q$. Often a statement of the form "$p \Rightarrow q$, and conversely" is used; this means the same as

$$(p \Rightarrow q) \land (q \Rightarrow p).$$

A second observation (one that you are asked to confirm as an exercise) is that the two statements

$$p \Rightarrow q \quad \text{and} \quad {\sim}q \Rightarrow {\sim}p$$

have the same value tables and thus are logically equivalent. The implication ${\sim}q \Rightarrow {\sim}p$ is called the *contrapositive* form of the implication $p \Rightarrow q$.

EXAMPLE 1 This example illustrates that correct reasoning from a false hypothesis can lead to statements that are false and to other statements that are true. Begin with the false arithmetic statement

$$5 = 1.$$

Add -3 to each side to obtain

$$5 - 3 = 1 - 3$$
$$2 = -2.$$

Now recall that if $a = b$ then $a^2 = b^2$, so

$$4 = 2(2) = (-2)(-2) = 4.$$

Hence, from the false statement $5 = 1$, we have used valid reasoning to deduce the false statement $2 = -2$, and the true statement $4 = 4$. Hence, if we begin with a false hypothesis, a statement that we derive by correct reasoning might be true or false. ∎

To explain what is meant by the term "correct reasoning," we need to return to the notion of a tautology, which by definition is a statement that has 1 in each position of its value table regardless of the values assigned to each of its component statements. That is, a tautology is *universally* true by virtue of its form as a statement. A few examples of tautologies are listed in the next theorem. To verify that a given statement is indeed a tautology, we can calculate the value table of that statement, or we can reduce the given statement algebraically to the form of a known tautology by applying properties of statements as given in Section 1.6. Each of these two methods is discussed in the examples that follow Theorem 1.7-1.

Theorem 1.7-1　Each of the following statements is a tautology. (Each statement has value 1 for every choice of p, q, and r.)

1. $p \Rightarrow p$.
2. $[(p \Rightarrow q) \wedge (q \Rightarrow p)] \Rightarrow (p \equiv q)$, and conversely.
3. $[(p \Rightarrow q) \wedge (q \Rightarrow r)] \Rightarrow (p \Rightarrow r)$.
4. $p \Rightarrow (p \vee q)$.
4*. $(p \wedge q) \Rightarrow p$.
5. $(p \Rightarrow q) \Rightarrow [(p \vee q) \equiv q]$.
5*. $(q \Rightarrow p) \Rightarrow [(p \wedge q) \equiv q]$.
6. $[(p \vee q) \equiv q] \Rightarrow (p \Rightarrow q)$.
6*. $[(p \wedge q) \equiv q] \Rightarrow (q \Rightarrow p)$.
13. $(p \Rightarrow q) \Rightarrow (\sim q \Rightarrow \sim p)$, and conversely.

Although the numbering of the statements in this theorem might seem haphazard, it should remind you of the theorems in Section 1.3. Look back to confirm that any property listed here can be obtained from the earlier property with the same number by replacing A by p, B by q, C by r, $\cup$ by $\vee$, $\cap$ by $\wedge$, $'$ by $\sim$, $\subseteq$ by $\Rightarrow$, and $=$ by $\equiv$. Therefore, in logic the implication symbol $\Rightarrow$ is analogous to the subset symbol $\subseteq$ of set theory. This conclusion demonstrates that the algebra of sets and the algebra of statements behave in the same way even though they differ in appearance.

The algebra of sets and the algebra of statements are examples of an algebraic system called a *boolean algebra*, in honor of George Boole (1815–1864), an Englishman who did much of the early work in formalizing the laws of thought. A contemporary mathematician, Marshall Stone, has proved that for any boolean algebra B there is a set S such that the algebra of some collection of subsets of S is exactly like that of B.

EXAMPLE 2　We now comment upon the first two tautologies of Theorem 1.7-1 and then prove the third.

(a) Property 1 can be established in two easy steps.

$$(p \Rightarrow p) \equiv p \lor {\sim} p \qquad \text{by Table 1.7-1}$$
$$\equiv u, \qquad \text{a tautology,}$$

by a statement preceding Example 2 in Section 1.6.

(b) Property 2 can be expressed more concisely by introducing a new symbol. If r and s are two statements, then we shorten the statement

$$(r \Rightarrow s) \land (s \Rightarrow r)$$

to

$$r \Leftrightarrow s,$$

read as "r if and only if s." (Recall that $r \Rightarrow s$ can be expressed as "r only if s," and $s \Rightarrow r$ can be expressed as either "r if s" or "if s then r.") Thus property 2 may now be written as

$$(p \Leftrightarrow q) \Rightarrow (p \equiv q) \quad \text{and} \quad (p \equiv q) \Rightarrow (p \Leftrightarrow q),$$

or even more concisely, because of the word "and," as

$$(p \Leftrightarrow q) \Leftrightarrow (p \equiv q).$$

Hence property 2 asserts that logical equivalence of two statements (defined in Section 1.6 as meaning identical value tables) means that each of the two statements implies the other.

(c) Property 3 states that implication is transitive. It can be proved by first replacing each $\Rightarrow$ by its equivalent expression using $\lor$ and $\sim$, and then by using the De Morgan laws and other properties from Section 1.6.

$$[(p \Rightarrow q) \land (q \Rightarrow r)] \Rightarrow (p \Rightarrow r)$$
$$\equiv [(q \lor {\sim} p) \land (r \lor {\sim} q)] \Rightarrow (r \lor {\sim} p)$$
$$\equiv (r \lor {\sim} p) \lor \{{\sim}[(q \lor {\sim} p) \land (r \lor {\sim} q)]\}$$
$$\equiv (r \lor {\sim} p) \lor \{{\sim}(q \lor {\sim} p) \lor {\sim}(r \lor {\sim} q)\}$$
$$\equiv (r \lor {\sim} p) \lor [({\sim} q \land {\sim}({\sim} p)) \lor ({\sim} r \land {\sim}({\sim} q)]$$
$$\equiv [r \lor {\sim} p \lor ({\sim} q \land p) \lor ({\sim} r \land q)]$$
$$\equiv [r \lor ({\sim} r \land q)] \lor [{\sim} p \lor (p \land {\sim} q)]$$
$$\equiv [(r \lor {\sim} r) \land (r \lor q)] \lor [({\sim} p \lor p) \land ({\sim} p \lor {\sim} q)]$$
$$\qquad\qquad\qquad\qquad\qquad\qquad \text{by distributivity}$$
$$\equiv [u \land (r \lor q)] \lor [u \land ({\sim} p \lor {\sim} q)]$$
$$\equiv (r \lor q) \lor ({\sim} p \lor {\sim} q) \equiv r \lor {\sim} p \lor (q \lor {\sim} q)$$
$$\equiv (r \lor {\sim} p) \lor u \equiv u, \qquad \text{a tautology.}$$

A value-table proof would have been quicker and easier, but our purpose was to practice using the symbols and properties of the algebra of statements.   ■

Now we are ready to comment on theorems and proofs in mathematics. A *theorem* is a statement of the form

$$p \Rightarrow q \qquad \text{(if } p \text{ then } q\text{)}$$

for which a proof has been given. Actually some theorems are stated in the form

$$p \Leftrightarrow q \qquad (p \text{ if and only if } q),$$

which, as we have seen, can be regarded as a pair of theorems, consisting of a theorem, $p \Rightarrow q$, and its converse, $q \Rightarrow p$.

A *proof* for our purposes will be any one of three forms of argument: (1) direct, (2) indirect, (3) inductive. Indirect proofs will be explained and illustrated in Section 1.8, and a description of inductive proofs will be given in Section 1.10.

A *direct proof* of the assertion $p \Rightarrow q$ is a finite sequence of statements, the first of which is p and the last of which is q, with intervening statements arranged in such a way that each statement is obtained from one or more preceding statements by applying a *rule of inference*.

To understand this definition, we need to know what is meant by an "inference" and by a "rule of inference." First we must recognize that the statement $p \Rightarrow q$ is made within a particular mathematical context and must be interpreted within that context; for example, we might be considering integer arithmetic, Euclidean plane geometry, probability, calculus, matrix algebra, or complex analysis. Each such context includes a body of statements that are considered to be true within that context. Initially a deductive system S begins with a list of statements that are declared to be true in S. Such statements are called *axioms*. No claim is made about the truth or falsity of those axioms in any universal sense, only that they are assigned the value 1 within S.

We previously described a theorem to be a statement of the form $p \Rightarrow q$ for which a proof has been given, and we described a direct proof to be a finite sequence of statements of the form

$$p$$
$$s_1$$
$$s_2$$
$$\cdot$$
$$\cdot$$
$$\cdot$$
$$s_k$$
$$q,$$

where each statement after p is obtained by applying a rule of inference. That means that each statement from s_1 through q is one of the following types:

1. An axiom of S
2. A theorem of S
3. A tautology
4. A statement r that has the value 1 whenever the value of s is 1, where s is a statement of the form $p \wedge s_1 \wedge s_2 \wedge \cdots \wedge s_j$, and each s_i has been introduced previously in this proof in one of these four ways

A statement r of type 4 is called an *inference* or a *deduction* from the *premises* $p, s_1, s_2, \ldots, s_j$. An inference is characterized completely by the condition that the value of r must be 1 whenever each of the premises is assigned the value 1, which is the same as saying that the statement

$$[p \wedge s_1 \wedge s_2 \wedge \cdots \wedge s_j] \Rightarrow r$$

has value 1 for any assignment of values to each of the premises. This can be written more simply as

$$s \Rightarrow r.$$

Note that s will have value 1 if and only if each of $p, s_1, \ldots, s_j$ has value 1, in which case r has value 1 by the description of type 4. Finally, we are able to deduce r from the pair of statements

$$s, \quad (s \Rightarrow r)$$

by applying one of the most frequently used rules of inference, called the *law of detachment* (or *modus ponens*):

If s and $(s \Rightarrow r)$ are true statements, then r is a true statement.

This law is justified by the fact that the statement

$$[s \wedge (s \Rightarrow r)] \Rightarrow r$$

is a tautology. Indeed, the same reasoning can be applied to obtain a rule of inference from any tautology that has the form of an implication (or logical equivalence). For example, tautology 5 of Theorem 1.7-1 yields this rule of inference:

If $p \Rightarrow q$ is a true statement, then so is $(p \vee q) \equiv q$.

Because tautologies play such an important role in deduction, we list here (for ease of reference in writing proofs) a few of the most frequently used rules of inference. Read them carefully, and think about what they say.

Some Rules of Inference   Each rule of inference is of the following form: "If the premise is true, so is the conclusion."

Premise	*Conclusion*
1. $p \wedge (p \Rightarrow q)$	q
2. $(p \Rightarrow r) \wedge (r \Rightarrow q)$	$p \Rightarrow q$
3. $(p \Rightarrow q) \wedge \sim q$	$\sim p$
4. $(p \vee q) \wedge \sim p$	q
5. $(p \vee q) \wedge (p \Rightarrow s) \wedge (q \Rightarrow t)$	$s \vee t$
6. $p \wedge q$	p
7. p	$p \vee q$
8. $(p \vee q) \Rightarrow r$	$(p \Rightarrow r) \wedge (q \Rightarrow r)$
8C. $(p \Rightarrow r) \wedge (q \Rightarrow r)$	$(p \vee q) \Rightarrow r$
9. $p \Rightarrow (q \wedge r)$	$(p \Rightarrow q) \wedge (p \Rightarrow r)$
9C. $(p \Rightarrow q) \wedge (p \Rightarrow r)$	$p \Rightarrow (q \wedge r)$
10. $\sim (p \Rightarrow q)$	$p \wedge \sim q$
10C. $p \wedge \sim q$	$\sim (p \Rightarrow q)$

Rule 8C is the converse of Rule 8; likewise for 9 and 9C and 10 and 10C.

EXAMPLE 3   To illustrate the method of direct proof, here is a formal proof of the following property of sets, called the *Modular Law*:
If A is a subset of C, then $A \cup (B \cap C) = (A \cup B) \cap C$.

1. $A \subseteq C.$ Hypothesis.
2. $A \cup (B \cap C) = (A \cup B) \cap (A \cup C).$ Part 9 of Theorem 1.3-2.
3. $A \subseteq C \Rightarrow [A \cup C = C].$ Part 5 of Theorem 1.3-1.
4. $A \cup C = C.$ Rule of Inference 1.
5. $A \cup (B \cap C) = (A \cup B) \cap C.$ Substitute 4 in 2.

Most proofs are written less formally, of course.   ■

Exercises 1.7

1. Construct value tables for each of the following:
    (a) $(q \lor \sim p) \land (\sim q \lor p).$
    (b) $p \Leftrightarrow q.$
    (c) $(p \lor \sim q) \Rightarrow \sim p.$
    (d) $\sim p \Leftrightarrow (p \Rightarrow \sim q).$
    (e) $(p \land \sim p) \Rightarrow q.$
    (f) the statements $p \Rightarrow q; q \Rightarrow p; \sim q \Rightarrow \sim p.$

2. State the contrapositive and the converse of each of these implications.
    (a) $(p \lor \sim q) \Rightarrow \sim p.$
    (b) $p \Rightarrow (q \Rightarrow r).$

3. Translate into words the following assertions of Theorem 1.7-1.
    (a) tautology 4: $p \Rightarrow (p \lor q).$
    (b) tautology 5: $(p \Rightarrow q) \Rightarrow [(p \lor q) \equiv q].$
    (c) tautology 6: $[(p \lor q) \equiv q] \Rightarrow (p \Rightarrow q).$
    (d) tautology 13: $(p \Rightarrow q) \Rightarrow (\sim q \Rightarrow \sim p)$, and conversely.

    Note that tautology 13 asserts that any implication and its contrapositive are logically equivalent statements.

4. Use value tables to prove the following statements of Theorem 1.7-1.
    (a) tautology 5*: $(q \Rightarrow p) \Rightarrow [(p \land q) \equiv q].$
    (b) tautology 13: $(p \Rightarrow q) \Rightarrow (\sim q \Rightarrow \sim p)$, and conversely.
    (c) tautology 4: $p \Rightarrow (p \lor q).$

5. Use the algebra of statements to prove each of the following statements of Theorem 1.7-1. You may use (c) of Example 2 as a general model.
    (a) tautology 4*: $(p \land q) \Rightarrow p.$
    (b) tautology 6*: $[(p \land q) \equiv q] \Rightarrow (q \Rightarrow p).$

6. Let the symbols f, b, and n, respectively, denote the statements: I go to France, I go to Belgium, I go to Norway. Use those letters to write each

statement below in symbolic form. Then simplify the symbolic statement and finally translate the simplified statement back into English.

(a) I will go to France only if I go to Belgium; and if I go to Norway, then I will go to France.

(b) If I do not go to France, then I will go to Belgium; and I will not go to France only if I go to Norway.

(c) If I go to France, then I will not go to Norway only if I do go to Belgium.

7. For each of the following statements write the converse and the contrapositive. Of the three statements (the given statement, its converse, and its contrapositive), which are true and which are false?

(a) If an integer is the square of some integer, then it is the fourth power of some integer.

(b) If the product of two positive integers is an even integer, then at least one of the two positive integers is even.

(c) If the product of two integers is positive, then each of the two integers is negative.

(d) If the sum of two positive integers is 20, then at least one of the integers is greater than 9.

(e) If the sum of two positive integers is even, then each of the two positive integers is even.

1.8   Indirect Proofs

This section describes and illustrates indirect proofs and demonstrates how the algebra of statements can be represented by electronic devices. Such devices enable computers to make complex logical decisions rapidly and reliably.

The form of an indirect proof of the statement $p \Rightarrow q$ is now easy to describe. From tautology 13 of Theorem 1.7-1 we know that $p \Rightarrow q$ is logically equivalent to its contrapositive $\sim q \Rightarrow \sim p$.

An *indirect proof* of the assertion $p \Rightarrow q$ is a direct proof of the contrapositive assertion $\sim q \Rightarrow \sim p$.

Hence an indirect proof of $p \Rightarrow q$ has the form of a direct proof in which the first statement is $\sim q$ (the negation of the conclusion q) and the last statement is $\sim p$ (the negation of the hypothesis p).

EXAMPLE 1   Here is an indirect proof of the following theorem of arithmetic:

> **Theorem** Let x and y denote natural numbers. If $x + y$ is odd, then either x is odd or y is odd.

Before we write an indirect proof, it will help to write the theorem symbolically:

$$p \Rightarrow (q \vee r)$$

where p denotes the statement, "$x + y$ is odd," q the statement, "x is odd," and r the statement, "y is odd." In an indirect proof we seek to establish the implication

$$\sim(q \vee r) \Rightarrow \sim p.$$

By one of the De Morgan laws that implication becomes

$$(\sim q \wedge \sim r) \Rightarrow \sim p.$$

In words, if x is not odd and y is not odd, then $x + y$ is not odd. But any natural number that is not odd must be even, so the implication can be restated as

"If x and y are both even, then $x + y$ is even."

Here is an indirect proof of the given theorem.

> **Proof** Let x and y be natural numbers such that $x + y$ is odd. To prove the conclusion "x is odd or y is odd" indirectly, assume the negation of that conclusion, namely, that x and y are both even. We seek to deduce the negation of the hypothesis p:
>
> $\sim p$:   "it is not the case that $x + y$ is odd,"
>
> or more simply, "$x + y$ is even."
>   Because x is even, x is divisible by 2. This means that $x = 2n$ for some natural number n. Likewise, because y is even, $y = 2m$ for some natural number m. Hence, we can write the equation
>
> $$x + y = 2n + 2m = 2(n + m).$$

Because $n + m$ is a natural number, the preceding sentence states that $x + y$ is divisible by 2, and hence $x + y$ is even. Thus $x + y$ is not odd. We have deduced the negation of the hypothesis p, and the proof is complete.   ∎

An indirect proof of the implication $p \Rightarrow q$ is called a *proof by contradiction,* because by assuming that the negation $(\sim q)$ of the conclusion q is true, we can deduce that the negation $(\sim p)$ of the hypothesis p is also true. Hence, if p is true and if q is not true, we can deduce the statement

$$p \wedge \sim p,$$

which is a contradiction. That is equivalent to saying that the statement $(p \wedge \sim q)$ is false, so its negation $\sim (p \wedge \sim q)$ is true. But then we also note that

$$\sim (p \wedge \sim q) \equiv (\sim p \vee q) \equiv (p \Rightarrow q).$$

Historically, the development of formal proofs of mathematical statements is attributed (at least by western scholars) to Greek civilization, extending from Thales and Pythagoras in the sixth century B. C. through Plato, Aristotle, and Euclid three centuries later. In writing his 13 books of the *Elements,* Euclid arranged and summarized most of the mathematical knowledge of that time, including plane and solid geometry, arithmetic, and number theory. Both direct and indirect proofs are used in the *Elements,* and Latin translations refer to the method of indirect proofs as *reductio ad absurdum.* In Book IX Euclid included his indirect proof of this property of natural numbers.

The number of primes is not finite.

A proof similar to Euclid's is begun in the next example; you may write a complete proof as an exercise.

EXAMPLE 2   By an indirect proof show that the number of primes in N is infinite.

First rewrite this statement in the form of an implication, letting P denote a nonempty set of primes.

If P is the set of all prime numbers, then P is not finite.

The hypothesis is

P is the set of all prime numbers,

and the conclusion is

P is not finite.

We want to show that the negation of the conclusion implies the negation of the hypothesis. That is, we want to show that "*P* is finite" implies that "*P* is not the set of all prime numbers." So suppose that *P* is finite. Let *n* denote the total number of primes in *P*, and let $p_1, p_2, \ldots, p_n$ denote all the primes in *P*. Next let *M* denote the number obtained by adding 1 to the product of all the primes in *P*:

$$M = (p_1 p_2 \cdots p_n) + 1.$$

Recall also that each natural number greater than 1 is divisible by at least one prime number. From this information we can readily deduce that the set *P* is not the set of all prime numbers; this deduction completes the proof. (See Exercise 3.)   ■

An Application of Boolean Algebra: Logic Gates

We have seen that any proposition in the algebra of sets, in the algebra of statements, or in any other boolean algebra can be expressed in terms of the little words "or," "and," "not." Those words, in turn, can be represented by electronic devices that enable computers to make the calculations of boolean algebra.

The word "or" can be represented by a black box with two input terminals and one output terminal. Inside the box is a device that will deliver a signal (higher voltage) to the output terminal if and only if a signal arrives at *either* or *both* input terminals.

The word "and" can be represented by a box that also has two input terminals and one output terminal, but its internal mechanism will deliver a high voltage output only when *both* input terminals receive a higher voltage signal simultaneously.

Finally, the word "not" can be represented by a box with one input terminal and one output terminal. The output signal is the reverse of the input signal. The three types of boxes are shown in Figure 1.8-1.

If we represent a lower voltage symbolically by 0 and a higher voltage by 1, then the set of outputs of a given circuit for all possible assignments of values to the inputs defines a value table for the statement that the circuit represents.

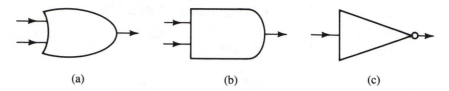

(a)   (b)   (c)

Figure 1.8-1 (a) OR gate; (b) AND gate; (c) NOT gate.

EXAMPLE 3   The statement

$$(p \text{ or } q) \text{ and } (\text{not } p)$$

can be modeled electrically by the circuit in Figure 1.8-2. The input-output values are given in Table 1.8-1.

But by using the algebra of statements we can write

$$(p \vee q) \wedge \sim p \equiv (p \wedge \sim p) \vee (q \wedge \sim p)$$
$$\equiv z \vee (q \wedge \sim p)$$
$$\equiv q \wedge \sim p.$$

Hence the original statement means precisely the same as the simpler statement

$$(q \wedge \sim p).$$

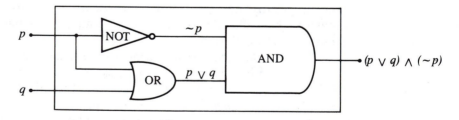

Figure 1.8-2

| Inputs | | Interior | | Output |
p	q	p or q	not p	(p or q) and (not p)
1	1	1	0	0
1	0	1	0	0
0	1	1	1	1
0	0	0	1	0

Table 1.8-1

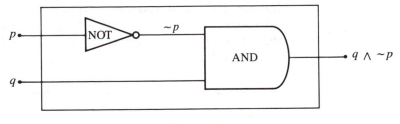

Figure 1.8-3

This means that the circuit representation can be constructed with only two devices inside the black box, resulting in potentially important improvements in manufacturing costs and product reliability. See Figure 1.8-3. ∎

Exercises 1.8

1. Using Example 1 as a guide, construct an indirect proof of the following statement about all pairs of natural numbers m and n: If mn is odd, then both m and n are odd.

2. Use an indirect proof to prove that if n is a natural number such that n^k is odd for some natural number $k > 0$, then n is odd.

3. Complete the discussion of Example 2 to prove that the number of primes is not finite.

4. Let m and n denote natural numbers. Use an indirect proof to show that if $m + n \geq 99$, then $m \geq 49$ or $n \geq 49$. (You may assume the truth of the following statement: If $a \leq b$ and $c \leq d$, then $a + c \leq b + d$.)

5. Use the drawings for the OR, AND, NOT devices to draw a circuit representation of each of the following statements.

    (a) $(\sim p \wedge \sim q) \vee p$.
    (b) $[(p \vee \sim q) \vee q] \wedge \sim p$.

6. Use the algebra of statements to simplify the statement in Exercise 5a in such a way that the circuit representation of the simplified statement has only two devices. Draw the circuit for the simplified statement.

7. Example 3 has only two propositions, p and q, corresponding to inputs. There is no reason why this number cannot be greater than two. Furthermore, to avoid crossing of wires and to make it easier to

interpret the circuit, you may wish to list some of the inputs more than once at the left end of the black box. To illustrate this, consider the following circuit representation of the statement $[(p \lor q) \land (\sim p \lor \sim q)] \lor r$.

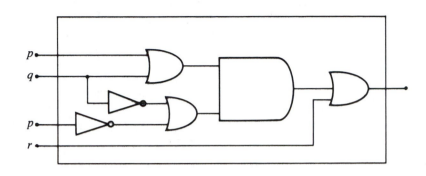

Determine the output signals of this circuit for the following input signals.

	Inputs	
p	*q*	*r*
1	1	1
1	1	0
0	1	0
0	0	1

8. Use the drawings for the OR, AND, NOT devices to draw a circuit representation of each of the following statements.

   (a) $p \land (\sim q \lor r)$.

   (b) $[(\sim p \land q) \lor r] \land \sim p$.

   (c) $\sim \{[(p \land \sim q) \lor r] \land p\} \lor (p \land q)$.

9. Use the algebra of statements to simplify the statement in Exercise 8c in such a way that the circuit representation of the simplified statement has only two devices. Draw the circuit for the simplified statement.

10. Express in terms of logical symbols the statements represented by the schematic diagrams of circuits shown on the facing page.

11. Use the fewest possible AND, OR, NOT devices to design a NOR circuit that has two inputs p, q and one output "neither p nor q." Explain why your circuit will produce the desired output.

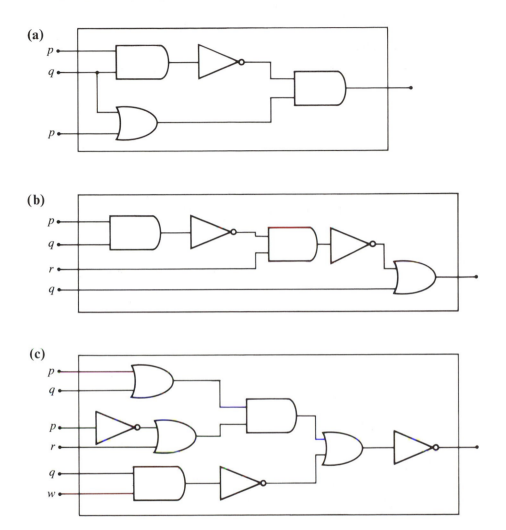

1.9 The Natural Numbers

Having developed an algebra of sets and an algebra of statements, we now consider some algebraic structures that are more familiar, namely, algebras of numbers. Mathematics makes use of many different number systems, including these that you have used in your prior studies.

$N =$ the set of natural numbers

$I =$ the set of integers

$Q =$ the set of rational numbers

$R =$ the set of real numbers

$C =$ the set of complex numbers

These sets of numbers are nested in the sense that within C each set is a subset of the next set:

$$N \subseteq I \subseteq Q \subseteq R \subseteq C,$$

and the arithmetic of each system is carried intact into the next system. For example, the complex number system C contains a subset of numbers that is an algebraic clone of the system R of real numbers.

Discrete mathematics makes use almost exclusively of the simplest of these number systems, N, which consists of all the numbers needed for counting finite sets:

$$N = \{0, 1, 2, 3, \ldots\}.$$

(Some mathematicians exclude zero from the set of natural numbers; inasmuch as the empty set is a finite set, we include zero in N, the counting numbers.) Occasionally we might need to use integers and rational numbers. The integers augment the counting numbers by providing an *additive inverse* (or *opposite*), $-m$, for each nonzero natural number m, with the property that $m + (-m) = 0$. Thus

$$I = \{0, 1, -1, 2, -2, \ldots\}.$$

Similarly, the rational numbers augment the integers by providing a *multiplicative inverse* (or *reciprocal*) for each nonzero integer k:

$$k \left(\frac{1}{k}\right) = 1 \qquad \text{for each } k \neq 0.$$

The word "rational" is used as the adjectival form of the word "ratio," which reflects the way we write the product of an integer j and the reciprocal of a nonzero integer k:

$$j \left(\frac{1}{k}\right) = \frac{j}{k} \qquad \text{for each } j \text{ and each } k \neq 0.$$

Thus any rational number is the ratio or *quotient* of two integers such that the *integer in the denominator is nonzero*. The symbol Q with which we designate the set of rational numbers reminds us that each rational number is a quotient of two integers.

Let us assume a familiarity with the arithmetic properties of these number systems. There are only two basic operations in the arithmetic of N: addition and multiplication. This is because the difference and the quotient of two natural numbers are not necessarily natural numbers;

$$3 \in N \quad \text{and} \quad 7 \in N, \quad \text{but} \quad 3 - 7 \notin N \quad \text{and} \quad \frac{3}{7} \notin N.$$

Hence we say that N is *not closed* under the operation of subtraction nor under the operation of division. The integers are closed under subtraction as well as under addition and multiplication, but they are not closed under division because

$$7 \in I \quad \text{and} \quad (-3) \in I, \quad \text{but} \quad \frac{-3}{7} \notin I.$$

But the rational number system Q is closed under the four usual operations of arithmetic, with the understanding that division by 0 is not defined, nor can it be defined without violating some customary and desirable property of arithmetic.

Decimal and Binary Representations of Numbers

The decimal representation of numbers is so familiar that we tend to forget what it means. We know that "thirty-five" is the name of the first natural number after the number whose name is "thirty-four," and the digital symbol we use in place of the word is 35 in decimal representation. This numerical symbol is a shortened way of writing

$$3(10)^1 + 5(10)^0.$$

More generally, the symbol 10435.62 is a shortened way of writing

$$1(10)^4 + 0(10)^3 + 4(10)^2 + 3(10)^1 + 5(10)^0 + 6(10)^{-1} + 2(10)^{-2}.$$

Of course, "10" is the number of distinct numerical digits that we have adopted for purposes of enumeration. Other than the genetic characteristic that gives most people 10 digits on their hands (usually five on each), there is no reason to prefer a system that uses 10 different numerical digits over a system that uses some other number of distinct numerical digits. Indeed, we have already seen situations in which two digits (0 and 1) appear to be a natural choice:

true/false

high/low voltage.

In electronic calculators and computers *binary representation* of rational numbers is used because the microdevices in contemporary electronics typically have two possible states, conducting or nonconducting, analogous to a light switch. The form of binary representation of a rational number is

$$\cdots + d_2(2)^2 + d_1(2)^1 + d_0(2)^0 + d_{-1}(2)^{-1} + d_{-2}(2)^{-2} + \cdots$$

with each coefficient d_k being 0 or 1, and only a finite number of nonzero coefficients of terms have positive exponents. Design limitations permit calculators and computers to have only a fixed number of coefficients altogether, which is an intrinsic source of error in machine computation.

EXAMPLE 1 Consider the following three numbers and their binary representations.

$$\begin{array}{rl} \text{Five:} & 1(2)^2 + 0(2)^1 + 1(2)^0 \quad \text{or } 101 \\ \text{Three-fourths:} & 0(2)^0 + 1(2)^{-1} + 1(2)^{-2} \quad \text{or } 0.11 \\ \text{Thirty-five:} & 1(2)^5 + 0(2)^4 + 0(2)^3 + 0(2)^2 + 1(2)^1 + 1(2)^0 \\ & \qquad\qquad\qquad\qquad\qquad \text{or } 100011 \end{array}$$

∎

Addition and multiplication of digits in binary notation are carried out very easily by learning the following tables.

+	0	1
0	0	1
1	1	10

×	0	1
0	0	0
1	0	1

For larger numbers we can add and multiply position by position as we did for decimal numbers.

EXAMPLE 2 The following two arithmetic problems appear in decimal and binary form.

$$\begin{array}{r} 35 \\ +27 \\ \hline 62 \end{array} \qquad \begin{array}{r} 100011 \\ +\ 11011 \\ \hline 111110 \end{array}$$

$$
\begin{array}{r}
35 \\
\times\ 27 \\
\hline
245 \\
70 \\
\hline
945
\end{array}
\qquad
\begin{array}{r}
100011 \\
\times\ 11011 \\
\hline
100011 \\
100011 \\
000000 \\
100011 \\
100011 \\
\hline
1110110001
\end{array}
$$

This last product is

$$1(2)^9 + 1(2)^8 + 1(2)^7 + 0(2)^6 + 1(2)^5 + 1(2)^4 + 0(2)^3 + 0(2)^2$$
$$+ 0(2)^1 + 1(2)^0$$

or

$$512 + 256 + 128 + 0 + 32 + 16 + 0 + 0 + 0 + 1 = 945. \qquad \blacksquare$$

EXAMPLE 3 This example describes a systematic method for obtaining the binary representation of any natural number n that is expressed in decimal form. The first step is to divide n by 2 and to write the quotient and the remainder. Each successive step consists of dividing the previous quotient by 2, recording the quotient and remainder, but stopping after completing the step in which the quotient first becomes 0. Finally the sequence of remainders, written in the *reverse order* from that in which they originally appeared, produces the binary representation of n. Here is an illustration with $n = 43$.

	Quotient	**Remainder**
$\dfrac{43}{2}$	21	1
$\dfrac{21}{2}$	10	1
$\dfrac{10}{2}$	5	0
$\dfrac{5}{2}$	2	1
$\dfrac{2}{2}$	1	0
$\dfrac{1}{2}$	0	1

The binary representation of 43 is 101011. $\qquad \blacksquare$

A computational scheme, such as the method described in Example 3, that solves each problem of a particular type in a finite number of steps is called an *algorithm*. The word is derived from the name of an Arabian scholar, al-Khowârizmî, whose ninth-century book on solutions of equations using Hindu numerals influenced European scholars for several centuries. The speed and accuracy of computers in carrying out the details specified in an algorithm has generated great interest in finding efficient algorithms that use fewer steps and thus less computer time.

Order Properties of *N*

Both addition and multiplication are ways in which we can combine a pair of numbers in *N* to obtain a third number in *N*. Each is therefore a *binary operation* on *N*, just as union and intersection are binary operations on the family of all subsets of a given set. The subset concept provides a way of relating certain pairs of sets to each other, so it is called a *binary relation* on a family of sets. The subset relation is a special instance of a *partial order relation,* a notion that is described below and explored further in Section 2.5. We now see how a similar but stronger type of order relation can be defined on the natural numbers.

> Let a and b be natural numbers: a is said to be *less than b*, written $a < b$, if and only if $a + d = b$ for some nonzero natural number d.

Let $a = 0$ and $b \neq 0$. Then $0 + b = b$, so $0 < b$. Every nonzero b in *N* is said to be *positive*.

Inasmuch as we assume that the arithmetic of *N* is familiar, we simply list the properties of the "less than" relation $<$, valid for all a, b, c in *N*.

1. Irreflexive: $a \not< a$.
2. Asymmetric: If $a < b$, then $b \not< a$.
3. Transitive: If $a < b$ and $b < c$, then $a < c$.
4. Trichotomy: Exactly one of the following is valid: $a < b$; $a = b$; $b < a$.

Any relation that satisfies the first three properties is called a *strong partial ordering*. When the fourth property also holds, the relation is called a *strong total ordering*. Thus, the relation $<$ describes a strong total ordering of *N*. Furthermore, this relation is preserved under both addition and multiplication, as described in the following two statements.

5. If $a < b$, then $a + c < b + c$.
6. If $a < b$, then $ac < bc$ if $c \neq 0$.

All the first five properties of strong order apply also to the "less than" relation defined on each of the number systems I, Q, and R. Because each of these systems contains negative numbers, however, the sixth property should be learned in more general form:

$$\text{If } a < b, \text{ then } \quad ac < bc \quad \text{if } c \text{ is positive,}$$
$$\text{but} \quad ac > bc \quad \text{if } c \text{ is negative,}$$
$$\text{and} \quad ac = bc \quad \text{if } c \text{ is zero.}$$

That is, multiplication of each side of an inequality by a negative number reverses the inequality. Of course, $ac > bc$ means the same as $bc < ac$.

There is another distinctive property of the ordering of N, a property that can be revised reasonably to apply in modified form in I but is not sensible in relation to the usual orderings of Q and R. The property that we refer to has a rather strange name — *well-ordering* — and a deceptively simple statement:

7. Every nonempty set of natural numbers has a smallest member. That is, if $S \subseteq N$ and $S \neq \varnothing$, then there exists $m \in S$ such that

$$m < a \text{ or } m = a \qquad \text{for each } a \in S.$$

EXAMPLE 4   (a) Suppose we need to refer to the first prime number larger than 10^{100}. Because there are infinitely many primes, the set S of all primes greater than 10^{100} must be nonempty. By the well-ordering property of N, S must have a smallest member, so the existence of the first prime larger than 10^{100} is firmly established even if we do not know its value.

(b) To see that the well-ordering property does not apply in I, let S be the set of all integers less than 10^{100}. Then S is nonempty, but certainly S does not have a smallest number. No matter what $x \in S$ we choose, the number $x - 1$ is also a member of S and is smaller than x. So the "smallest integer less than 10^{100}" is a fallacious concept in I.

(c) Similarly, "the smallest positive rational number" is a fallacious concept. Try to explain this assertion, first to yourself and then to a classmate.   ∎

In Section 1.10 we shall see how the well-ordering property of N, which seems so intuitively clear, can be used to justify the third method of proof listed in Section 1.7.

A modified form of the well-ordering property, stated below, also applies to any set of integers that has a *lower bound b* in I:

7′. Let S be a set of integers such that for some integer b

$$b < s \quad \text{for every } s \in S.$$

Then every nonempty subset of S has a smallest member.

From the well-ordering property we are led to suspect that a significant difference among the number systems N, I, Q, and R can be seen from the manner in which the members of each system are distributed according to order along a coordinate axis, as illustrated in Figure 1.9-1.

The points of N occur at equally spaced positions along the line, starting at the point labeled zero and moving to the right step by step. The points of I include those of N and also the mirror images of all of the points of N, extending from zero to the left along the line at one-unit intervals. The points of both N and I are discretely spaced along the line. But on Q the points are closely packed together, and they are even more so on R. To see this, suppose that $a < b$ in Q. The point on the line corresponding to the rational number a lies to the left of the point that corresponds to b. Since a and b are rational numbers, so is the number

$$h = \frac{a + b}{2}.$$

(Can you prove that $h \in Q$?) The point corresponding to h is halfway between a and b. So we now have three distinct rational numbers

$$a < h < b.$$

We can repeat the process using a and h in place of a and b, getting another rational number k, the midpoint of the segment from a to h. The process can be repeated time after time, without end:

Between any two distinct rational numbers there are *infinitely many* other rational numbers.

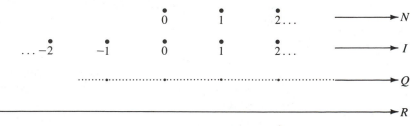

Figure 1.9-1

Hence the rational points are spread thickly and uniformly along the axis. But not all points on the line correspond to rational points because there are positions that correspond to irrational real numbers, such as π and $\sqrt{2}$. Between any two distinct points on the line there are infinitely many points with rational coordinates and infinitely many points with irrational coordinates, and each point on the line corresponds to a uniquely determined real number.

In this section we have looked briefly at some familiar number systems, with emphasis on N and on order properties of N that are also order properties in I, Q, and R. In Section 1.10 we shall see that the well-ordering property provides a method of proof that is valid in N and (in a restricted form) in I but not in Q, R, or C.

Exercises 1.9

1. Express each of the following numbers in decimal notation.
   (a) $3(10)^3 + 4(10)^2 + 6(10)^0 + 5(10)^{-2} + 7(10)^{-3}$
   (b) $5(10)^{-2} + 7(10)^{-6}$

2. Express each of the following numbers in binary notation.
   (a) $1(2)^5 + 1(2)^2 + 1(2)^0 + 1(2)^{-3}$
   (b) $1(2)^{-4} + 1(2)^{-6}$

3. Express each of the integers from 1 through 10 in binary notation.

4. Convert each of the following decimal representations of numbers to binary form.
   (a) 31
   (b) 203
   (c) 32.25
   (d) 0.625

5. Convert each of the following binary representations of numbers to decimal form.
   (a) 1011
   (b) 10101
   (c) 0.101
   (d) 101.001

6. Use binary arithmetic to perform the following operations, where each of the numbers is given in binary form.
   (a) $111 + 1011$
   (b) $101101 + 111110$
   (c) 101×111
   (d) 11011×11001

Check your answers to Exercise 6 by translating your work into the decimal system.

7. Answer the following questions:
   (a) Use the algorithm of Example 3 to convert the decimal number 61 to binary form.
   (b) Use your work in **a** to demonstrate why the algorithm works for 61.

8. An n-bit computer uses a sequence of n binary digits (0s and 1s) to represent each number, with the first digit used to designate whether the number is positive (0) or negative (1). Thus, a 4-bit computer can contain at most 16 (that is, 2^4) integers. List all sequences for a 4-bit computer, and then convert each of these sequences to decimal form.

9. Answer the following questions:
   (a) Why can an integer be regarded as a rational number?
   (b) In the discussion concerning rational numbers, it was asserted without proof that the number $h = (a + b)/2$ is rational if a and b are rational. Prove this assertion.
   (c) Prove that if r and s are rational numbers, then so are $r + s$ and rs.

10. Explain why division by 0 cannot be defined in Q without violating some property of arithmetic.

11. Use an indirect proof and the well-ordering property of N to prove that no natural number m satisfies $0 < m < 1$.

12. Give a careful explanation of the assertion: "The smallest positive rational number is a fallacious concept."

13. Use an indirect proof to prove the following statement, where a and b are integers and $b \neq 0$: If $\sqrt{2}$ is not rational, then $a + b\sqrt{2}$ is not rational.

14. Use an indirect proof to show that the real number $\sqrt{2}$ is irrational (not rational).

15. Answer the following questions and justify your answers.
   (a) Can the sum of a rational number and an irrational number be rational?
   (b) Can the sum of two irrational numbers be rational?
   (c) Is the set of irrational numbers closed under addition? Under multiplication?

16. Suppose we think of arranging the elements of I in the following way:

$$0, 1, 2, 3, \ldots, -1, -2, -3, \ldots,$$

where, if we read from left to right, the natural numbers occur in their usual order, the negative integers occur in the reverse of their usual order, and each natural number precedes every negative integer.

For any two integers a and b let us write $a \lhd b$ if and only if a *appears to the left of b* in the above arrangement.

(a) Which of properties 1 to 4 of the usual order relation $<$ hold for the unusual relation denoted by $\lhd$? You need not prove that your answers are correct, but some explanation should be given.

(b) If $S \subseteq I$ and $S \neq \varnothing$, does there exist $m \in S$ such that for each $a \in S$ either $m \lhd a$ or $m = a$? Explain your answer.

Exercises 17 through 20 introduce a third way of representing rational numbers. Decimal representation is sometimes referred to as base 10 representation, and binary as base 2 representation. These four exercises use base 8 representation. In base 8 representation we use only the digits $0, 1, 2, \ldots, 7$, and we write

seventeen as 21 because $17 = 2(8)^1 + 1(8)^0$

one hundred seventy-five as 257 because $175 = 2(8)^2 + 5(8)^1 + 7(8)^0$

seventy-five hundredths as 0.6 because $0.75 = 0(8)^0 + 6(8)^{-1}$.

17. Convert each of the following base 10 representations of numbers to base 8 representation.

    (a) 93
    (b) 649
    (c) 41.25
    (d) 0.625

18. Convert each of the following base 8 representations of numbers to base 10 representation.

    (a) 54
    (b) 2067
    (c) 0.13
    (d) 7.24

19. Construct an addition table and a multiplication table for base 8 arithmetic.

20. Use base 8 arithmetic to perform the following operations, where each of the numbers is given in base 8 form.

    (a) $123 + 56$
    (b) $1436 + 465$
    (c) 12×16
    (d) 52×117

    Check your answers by translating your work into the decimal system.

1.10　Mathematical Induction

PROBLEM　**The Domino Theory**

This question is concerned not with the domino theory of world politics but with actual dominoes. Suppose you want to set a new world's record for the number of dominoes that can be placed on end individually, no domino touching any other domino, such that when the first domino is toppled it will topple the second, which in turn will topple the third, and so on until all the dominoes fall. And suppose you have available as many dominoes as there are natural numbers. For example, a large shipment of dominoes might arrive every day faster than you can set them up. After several months of setting up a long trail of dominoes, you are ready to challenge the world's record. But before you start the process of successive toppling, you want to make a last inspection of your work. What will you need to verify about the way the dominoes are placed to be certain that all can be made to fall by causing the first domino to fall?　■

If you can answer that question clearly and correctly, you are on your way toward understanding the mathematical concept of reasoning by induction.

Principle of Mathematical Induction　Let S be any nonempty subset of the set N of natural numbers. If

1. $s \in S$ for some specific s in N,

and if

2. $m + 1 \in S$ whenever $m \in S$,

then

3. S contains all natural numbers $k \geq s$.

The number s in statement 1 is called a *starting value*. Statement 2 then can be used to deduce that $s + 1 \in S$ because $s \in S$; $s + 2 \in S$ because $s + 1 \in S$; and so on for each larger natural number. Often the smallest starting value is 0 or 1.

This principle can be used to prove properties that turn out to be true for every natural number. If $P(n)$ is a statement about the natural number n, let S denote the set of all $k \in N$ for which $P(k)$ is true. The principle of mathematical induction then asserts that if $0 \in S$, and if $m + 1 \in S$ whenever $m \in S$, then $S = N$; that is, $P(k)$ is true for each $k \in N$.

Observe carefully the form of statements 1 and 2 that constitute the premises (hypotheses) of the induction principle. The first is a specific declaration that $P(s)$ is true. We call statement 1 the *starting step.* However, statement 2 is an implication: The premise of 2 is that $P(m)$ is true for some unspecified natural number m, and the conclusion of 2 is that $P(m + 1)$ must be true. We call statement 2 the *inductive step.* The premise of 2 is called the *induction hypothesis.* It is vital that you understand that the inductive step *does not assert* that $P(m)$ is true. And it does not assert that $P(m + 1)$ is true. It only asserts that if $P(m)$ is true for some m, then $P(m + 1)$ must also be true. The combination of statements 1 and 2 works like this:

$P(s)$ is true by statement 1.

Because $P(s)$ is true, $P(s + 1)$ must be true by statement 2.

Because $P(s + 1)$ is true, $P(s + 2)$ must be true by statement 2.

And so on, for every natural number $n \geq s$.

The inductive step resembles a perpetual motion machine that marches on, one step at a time, once it has been started. But it has no self-starter and must be set in motion by another mechanism, namely the starting step, statement 1. Together, statements 1 and 2 imply the *conclusion: P(k)* is valid for all $k \in N$ such that $k \geq s$.

EXAMPLE 1   Let us compute for a few values of m the total sum $t(m)$ of values of the first m odd natural numbers. The first few odd natural numbers are

$$1, 3, 5, 7, 9, \ldots, (2k - 1), (2k + 1), \ldots.$$

We assert that the mth odd natural number is $2m - 1$. It is an easy matter to prove that assertion by induction, which you may do as an exercise. So let's try to establish the value of $t(m)$ on the assumption that the mth odd natural number is $2m - 1$. We have

$$t(0) = 0,$$
$$t(1) = 0 + 1 = 1,$$
$$t(2) = 1 + 3 = 4,$$
$$t(3) = 4 + 5 = 9,$$
$$t(4) = 9 + 7 = 16,$$
$$t(5) = 16 + 9 = 25.$$

We have defined $t(0)$, the sum of the first zero odd natural numbers, to be zero so that the pattern of these calculations appears to be

$$t(m + 1) = t(m) + (2m + 1) \qquad \text{for all } m \in N.$$

Check to be sure you see why this equation holds. Also it appears likely that $t(k) = k^2$ for all $k \in N$.

Now let's write a proof by induction of this assertion: "The sum of the first k odd natural numbers is k^2." Let S denote the set of natural numbers k for which that assertion is valid.

Starting Step. Because the sum of the first zero odd natural numbers is 0, we have $t(0) = 0^2$, so $0 \in S$. Thus 0 is a starting value.

Inductive Step. The induction hypothesis is "suppose that $m \in S$"; then

$$t(m) = m^2$$

for an unspecified natural number m. Then $t(m + 1)$ is the sum of $t(m)$ and the $(m + 1)$st odd natural number:

$$
\begin{aligned}
t(m + 1) &= t(m) + [2(m + 1) - 1] \\
&= m^2 + 2m + 2 - 1 \qquad \text{(by the induction hypothesis)} \\
&= m^2 + 2m + 1 = (m + 1)^2.
\end{aligned}
$$

Hence $(m + 1) \in S$ whenever $m \in S$.

Conclusion. By the principle of induction the formula $t(m) = m^2$ is correct for every natural number m. That is, $t(0) = 0$, and

$$1 + 3 + 5 + \cdots + (2m - 1) = m^2 \qquad \text{for all } m > 0.$$

It is, of course, only the second equation that is of interest to us. ∎

EXAMPLE 2   Prove by induction that a k-sided regular polygon has $k(k - 3)/2$ interior diagonals when $k \geq 3$.

Starting Step. Recall that the k vertices of a k-sided regular polygon lie on the circle that circumscribes the polygon. If $k < 3$, no polygon is formed, and if $k = 3$, there are no interior diagonals, and $3(3 - 3)/2 = 0$. Hence the formula is correct for $k = 3$, and 3 is a starting value.

Inductive Step. For $m \geq 3$ assume that there are $m(m - 3)/2$ interior diagonals of an m-sided polygon. (This is the induction hypothesis.) Now consider a regular polygon with $m + 1$ sides and hence $m + 1$

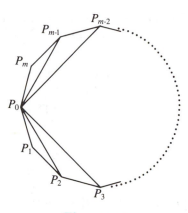

Figure 1.10-1

vertices. Pick any vertex, label it P_0, and label the other vertices P_1 to P_m around the circle from P_0. See Figure 1.10-1. Note that there are at least two other vertices, because $m \geq 3$; this is essential to the argument that follows.

For a diagonal of the polygon to be an interior diagonal, it cannot connect a pair of adjacent vertices, but any nonadjacent pair of vertices determines a diagonal that is interior to the polygon. If we ignore P_0 for the moment, there are m vertices and hence $m(m - 3)/2$ interior diagonals connecting pairs of nonadjacent pairs of $\{P_1, P_2, \ldots, P_m\}$. Now what is the effect of introducing P_0? The segments from P_0 to $P_2, \ldots, P_{m-1}$ are new interior diagonals, and there are $m - 2$ of these. Furthermore, P_1 and P_m are no longer adjacent, so that segment becomes interior after P_0 is introduced. No previously interior diagonals are lost by introducing P_0, so any $(m + 1)$-sided regular polygon has $(m - 2) + 1 = m - 1$ more interior diagonals than does an m-sided polygon. By the induction hypothesis this new total is

$$\frac{m(m - 3)}{2} + m - 1,$$

which can be reduced algebraically to

$$\frac{m^2 - 3m + 2(m - 1)}{2} = \frac{m^2 - m - 2}{2} = \frac{(m + 1)(m - 2)}{2}$$
$$= \frac{(m + 1)[(m + 1) - 3]}{2}.$$

Notice that this is the given formula for $k = m + 1$. Thus the formula is correct for $k = m + 1$ whenever it is correct for $k = m$.

Conclusion. By induction the number of interior diagonals of a regular polygon with k sides ($k > 2$) is $k(k - 3)/2$.   ∎

EXAMPLE 3   The Tower of Hanoi puzzle is played with disks of different sizes that can be stacked on three pegs. At the start of the puzzle all the disks are arranged on one of the pegs in such a way that each disk, except the bottom one, is smaller than the disk just below it. The object of the puzzle is to move the tower to either of the two other pegs by moving one disk at a time to any peg (subject to the constraint that no disk may be placed on top of a smaller disk) and to determine the smallest possible number of moves.

Let $H(n)$ denote the smallest number of moves required to solve the n-disk puzzle. It should be apparent that $H(1) = 1$. Figure 1.10-2 illustrates how a tower of two disks can be moved to one of the other two pegs in three steps, and $H(2) = 3$.

Because we want to determine $H(n)$ for all $n > 0$, we should try to discover a general method of moving the disks that makes use of prior experience. For example, we can solve the n-disk puzzle by the sequence of moves listed at the top of the facing page.

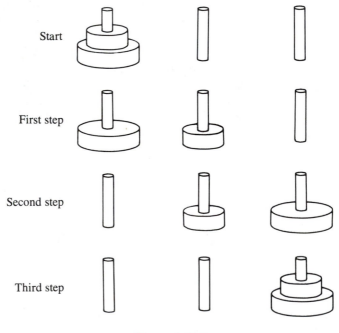

Figure 1.10-2

1. Transfer the top $n - 1$ disks from peg 1 to peg 2, using $H(n - 1)$ individual moves, ignoring completely the largest disk.

2. Transfer the largest disk from peg 1 to peg 3, using $H(1) = 1$ move.

3. Transfer the $n - 1$ disks from peg 2 to peg 3, using $H(n - 1)$ additional moves, thus completing the n-disk puzzle in $H(n)$ moves, where

$$H(n) = H(n - 1) + 1 + H(n - 1) = 2H(n - 1) + 1.$$

This equation tells us how $H(n)$ can be calculated from $H(n - 1)$, and we see the possibility of an inductive argument. Next we need a general formula for $H(n - 1)$. We can use the values of $H(1)$ and the general equation to calculate

$$H(1) = 1,$$
$$H(2) = 2(1) + 1 = 3,$$
$$H(3) = 2(3) + 1 = 7,$$
$$H(4) = 2(7) + 1 = 15,$$
$$H(5) = 2(15) + 1 = 31,$$

and so on. Now it is reasonably evident that the pattern for $H(k)$ is given by

$$H(k) = 2^k - 1.$$

To prove that this formula is correct, we write an inductive proof.

Starting Step. The 1-disk puzzle can be solved by one move, and $1 = 2^1 - 1$, so 1 is a starting value.

Inductive Step. Assume that the formula is valid for $k = m$. Then the m-disk tower can be transferred in $H(m) = 2^m - 1$ moves. Furthermore, as described previously, an $(m + 1)$-disk tower can be transferred in $2H(m) + 1$ moves; that is,

$$H(m + 1) = 2H(m) + 1 = 2(2^m) - 1 = 2^{m+1} - 1.$$

Thus the formula $H(k) = 2^k - 1$ is valid for $k = m + 1$ whenever it is valid for $k = m$.

Conclusion. By induction, for $n \geq 1$ the n-disk Tower of Hanoi puzzle can be solved with $2^n - 1$ individual moves. ∎

Next we explore the relation between well-ordering and induction and show that these two principles are logically equivalent. That means, of course, that the validity of either principle implies the validity of the other. That fact is worth remembering, because in some instances a proof using one of those two principles is easier to formulate than a proof using the other.

First we show that the well-ordering principle implies the principle of mathematical induction. Let S be any set that satisfies statement 1 ($s \in S$ for some $s \in N$) and statement 2 (whenever $m \in S$, then $m + 1 \in S$) of the principle of mathematical induction. We want to prove statement 3 (S contains all natural numbers $k \geq s$). Let T denote the set of all $t \in N$ such that $t \geq s$ but $t \notin S$. Suppose for the moment that T is nonempty. Because N is well-ordered, T then contains a smallest number k. By definition of T, $k \geq s$ and $k \notin S$. Hence $k > k - 1 \geq s$. But $k - 1 \notin T$ because k is the smallest number in T. So $k - 1 \in S$. By statement 2, we deduce that $k \in S$, which contradicts the fact that $S \cap T$ is empty. Hence, the temporary supposition that T is nonempty must be abandoned, which means that statement 3 is valid.

Now we consider the converse: The principle of mathematical induction implies the well-ordering principle. To prove that assertion it will suffice for us to prove by induction that each nonempty subset of N contains a smallest member.

Certainly every subset of N that contains 0 has 0 as a smallest member. Denote the family of all such sets by $F(0)$. Likewise any set A that contains 0 or 1 must have a least member. If $A \in F(0)$, that smallest member is 0; otherwise the smallest member of A is 1. Let $F(1)$ denote the family of all subsets of N that contain some number not exceeding 1. Then $F(0) \subseteq F(1)$, and each set in $F(1)$ has a smallest member. Now let $F(2)$ denote the family of all subsets of N that contain some number not exceeding 2. Then

$$F(0) \subseteq F(1) \subseteq F(2),$$

and each set belonging to $F(2)$ either has 2 as a smallest member or belongs to $F(1)$ and therefore has a smallest member.

It appears that such a family $F(n)$ of sets can be defined for each n. Furthermore, any nonempty subset A of N must contain some natural number, $m + 1$, and therefore must belong to the family $F(m + 1)$. Also any set in $F(m + 1)$ either has $m + 1$ as a smallest member or belongs to $F(m)$. If an inductive proof has

"Suppose each set in $F(m)$ has a smallest member"

as its induction hypothesis, then the same property is easily deduced for $F(m + 1)$ and, by induction for $F(n)$, for every $n \in N$.

The previous paragraphs describe how a proof by induction might be conceived and planned before it is written. You will not be surprised to find among the exercises a request that you write in detail such a proof.

Examples of Incorrect Reasoning Using Induction

To emphasize that an inductive proof must establish *both* statements 1 and 2 to correctly establish statement 3, we now consider some examples in which failure to establish statement 1 or statement 2 rigorously can lead to incorrect conclusions.

EXAMPLE 4 Determine a formula in terms of n for the sum of the squares of the first n odd positive integers and prove that the formula is valid for all n.

Suppose an algebraic mistake is made in writing a formula, and the following expression is proposed:

$$1^2 + 3^2 + \cdots + (2n - 1)^2 = \frac{8n^3 - 2n + 6}{6}.$$

(The correct formula differs from this only by having 0 in place of 6 in the numerator.) When this formula is used as the induction hypothesis, the inductive step produces the corresponding expression for the sum of the first $n + 1$ odd integers. That is, step 2 shows that if the formula is correct for any natural number n, then it is also correct for the next natural number, $n + 1$:

$$1^2 + 3^2 + \cdots + (2n - 1)^2 + (2n + 1)^2 = (2n + 1)^2 + \frac{8n^3 - 2n + 6}{6}$$

$$= \frac{8(n + 1)^3 - 2(n + 1) + 6}{6}.$$

But the formula itself is incorrect for every $n \in N$. This example is comparable to setting the dominoes with correct spacing and alignment but setting each one in concrete so that none can be toppled. It illustrates that step 1 of induction, even though it is often very easy to verify, is essential in an inductive proof. ■

EXAMPLE 5 Prove by induction that $b^n = 1$ for any number $b \neq 0$ and each natural number n.

Let S denote the set of all natural numbers n for which $b^n = 1$.

Starting Step. $b^0 = 1$ by definition, so $0 \in S$.

Inductive Step. Assume that $m \in S$, so $b^m = 1$. Then

$$b^{m+1} = \frac{b^m b^m}{b^{m-1}} = \frac{1(1)}{1} = 1,$$

so $m + 1 \in S$. By the principle of induction, $S = N$, so $b^n = 1$ for any $b \neq 0$ and every $n \in N$.

See whether you can spot the error of reasoning. One way to examine an argument that leads to a result that is clearly wrong is to work through the argument carefully, assigning specific values to the number symbols.

■

EXAMPLE 6 Prove by induction that there is no positive natural number n such that $991n^2 + 1$ is a perfect square (the square of some natural number).

Let S denote the set of all positive natural numbers n for which $991n^2 + 1$ is not a perfect square.

Starting Step. If $n = 1$, then 992 is not a perfect square because $31^2 = 961$ and $32^2 = 1024$. Hence $1 \in S$.

Inductive Step. Assume that $m \in S$. That is, assume that

$$991m^2 + 1 \neq x^2$$

for every natural number x. We'd like to use that information to show that $m + 1 \in S$, that is, to show that $[991(m + 1)^2 + 1]$ is not a square. We have

$$991(m + 1)^2 + 1 = 991(m^2 + 2m + 1) + 1$$
$$= (991m^2 + 1) + 991(2m + 1).$$

At this stage there is not any promising way to proceed with the algebra, and we might wonder if the result is really correct. So we might look for a counterexample—a specific value of n for which $991n^2 + 1$ is a perfect square. By this approach you could show that $2 \in S$, $3 \in S$, $4 \in S$, and so on. If you continued in this way, checking 1000 more numbers each day of your life, you wouldn't find a number not in S. But the 29-digit number written at the end of the first paragraph of Section 1.4 happens not to be in S, and yet every smaller positive integer is in S. The world of numbers has some surprises after all!   ■

Exercises 1.10

In all exercises where you use a proof by induction, label the *starting step,* the *inductive step,* and the *conclusion.* Also indicate where the induction hypothesis is used.

1. Use mathematical induction to prove that for all natural numbers k:
    (a) $0 + 1 + 2 + 3 + \cdots + k = [k(k + 1)]/2$.
    (b) If $k > 0$, the kth odd natural number is $2k - 1$.

2. Derive the formula proved in Exercise 1a by adding the following two equations and simplifying.

    $$t = 0 + \quad 1 + \quad 2 + \quad 3 + \cdots + (k - 1) + k.$$
    $$t = k + (k - 1) + (k - 2) + (k - 3) + \cdots + \quad 1 \ + 0.$$

3. Give a clearly stated answer to the question posed in the domino theory problem at the beginning of this section.

4. Let $P(n)$ denote the statement

    "The sum $2^0 + 2^{-1} + 2^{-2} + \cdots + 2^{-n}$ is equal to $2 - 2^{-n}$."

    Prove that $P(n)$ is true for all natural numbers n.

5. An incorrect formula for the sum of the squares of the first n odd positive integers was given in Example 5. The correct formula is

    $$1^2 + 3^2 + 5^2 + \cdots + (2n - 1)^2 = \frac{4n^3 - n}{3}.$$

    Use induction to prove that this formula is correct for all $n \geq 1$.

6. Write an inductive proof of the assertion, "The sum of the squares of the first n even natural numbers is $2(n - 1)(n)(2n - 1)/3$." (Remember that the first even natural number is zero.)

7. Prove that

    $$1^3 + 2^3 + 3^3 + \cdots + n^3 = \frac{n^2(n + 1)^2}{4}$$

    for every natural number $n \geq 1$.

8. In Exercise 9 of Section 1.4 you were asked to derive a formula for the number of subsets of an n-element set, where n is a positive integer. If

you did that exercise you probably concluded that there are 2^n subsets. Check this by using induction to prove that an n-element set has 2^n subsets.

9. Prove that the following assertion is true for every natural number k: If r is any real number such that $r > -1$, then

$$(1 + r)^k \geq 1 + kr.$$

10. Determine all natural numbers n for which $2^n > n^2$ and prove that your answer is correct.

11. Let k be any negative integer and let $W(k)$ be the set of all integers not less than k. Show that $W(k)$ is well-ordered by the "less than" relation.

12. The sequence of natural numbers

$$1, 1, 2, 3, 5, 8, 13, 21, \ldots,$$

where each entry after the second is the sum of the two preceding entries, is called a *Fibonacci sequence*. So if $F(k)$ denotes the kth term of this sequence, then $F(k + 1) = F(k) + F(k - 1)$ for all $k \geq 2$.

(a) Use the well-ordering principle to prove that $F(k) < 2^k$ for all positive integers k.
(b) Prove that $F(k) < (7/4)^k$ for all positive integers k.

13. Using the argument preceding Example 4 as a guide, write in full detail a proof that induction implies well-ordering in N.

14. Let $A_1, A_2, \ldots, A_k$ be finite sets such that $A_i \cap A_j = 0$ for all i and j with $i \neq j$.

(a) Show that the formula

$$n(A_1 \cup A_2 \cup \cdots \cup A_k) = n(A_1) + n(A_2) + \cdots + n(A_k)$$

follows from the general inclusion-exclusion principle as explained in Section 1.5.
(b) Prove by induction that the formula given in (a) is valid for all positive integers k.

15. Prove that

$$1(3) + 2(4) + 3(5) + \cdots + n(n + 2) = \frac{(n)(n + 1)(2n + 7)}{6}$$

for all natural numbers $n \geq 1$.

16. Prove that

$$\frac{1}{1(2)} + \frac{1}{2(3)} + \frac{1}{3(4)} + \cdots + \frac{1}{k(k+1)} = \frac{k}{k+1}$$

for all positive integers k.

17. Prove that

$$\frac{1}{1(3)} + \frac{1}{3(5)} + \frac{1}{5(7)} + \cdots + \frac{1}{(2k-1)(2k+1)} = \frac{k}{2k+1}$$

for all positive integers k.

18. What formula is suggested by Exercises 16 and 17 for the following sum?

$$\frac{1}{1(4)} + \frac{1}{4(7)} + \frac{1}{7(10)} + \cdots + \frac{1}{(3k-2)(3k+1)}$$

Use induction to try to prove that your guess is correct.

19. Prove (by induction or otherwise) that the number $13^n - 5^n$ is evenly divisible by 8 for all positive integers n.

20. Prove (by induction or otherwise) that the number $n^3 - n$ is evenly divisible by 3 for every positive integer n. (This assertion is a special case of a theorem of Fermat, which states that $n^p - n$ is evenly divisible by p if p is a prime.)

21. Example 1 proved that the sum $t(k)$ of the first k odd natural numbers is equal to k^2. A student is baffled, because according to the student's work the sum should be $k^2 + 5$. Here is the student's proof. Assume that

$$1 + 3 + 5 + \cdots + (2m - 1) = m^2 + 5.$$

Then

$$1 + 3 + 5 + \cdots + (2m - 1) + (2m + 1)$$
$$= [1 + 3 + 5 + \cdots + (2m - 1)] + (2m + 1)$$
$$= (m^2 + 5) + 2m + 1$$
$$= (m^2 + 2m + 1) + 5$$
$$= (m + 1)^2 + 5.$$

Hence, if $t(k) = k^2 + 5$ is true for $k = m$, it is true for $k = m + 1$. By induction, $t(k) = k^2 + 5$ for all natural numbers $k \geq 1$.
Explain the error in the student's proof.

Chapter 2

RELATIONS, FUNCTIONS, AND OPERATIONS

This chapter continues to lay foundations that are essential for the understanding and application of the mathematical sciences. Starting from the concept of cartesian product of sets, introduced in Section 1.4, we study binary relations and some other special types of relations that occur frequently, not only in mathematics but also throughout human experience. Then functions are introduced as special types of relations, and mathematical operations are described as special types of functions. In Section 2.7 all these concepts are combined with the material of Chapter 1 to explain what is meant by a "mathematical structure."

BASIC CONCEPTS

2.1 Binary Relations

College admission application forms frequently ask each applicant to list all the extracurricular activities in which the applicant has participated during high school. For a given school and a given graduating class it would be possible to make two lists: one containing the name of each class member, and the other itemizing the official extracurricular activities of the school.

Seniors	*Activities*
Abel, N.	Aquatic club
Bolyai, H.	Band
Cayley, E.	Cross-country team
Descartes, F.	Dramatics
Euler, S.	Equestrian team
Fibonacci, I.	French club
Gauss, G.	Gardening club
Hypatia, A.	Honor society

To show the activities of any senior, we could draw an arrow from that senior's name to each activity in which he or she participated. If this were done for every senior, then we could obtain a list of seniors involved in each activity simply by following back all the arrows from the activities to the seniors. If these arrows were drawn for even a moderately sized class, the page would look like a plate of spaghetti, so this is not a practical way to keep track of the information. However, it does represent quite accurately the concept of a binary relation.

The word "binary" means that two lists or sets (not necessarily distinct from each other) are involved. Each arrow relates a member of the first set A to a member of the second set B and therefore determines an ordered pair (a, b), where $a \in A$ and $b \in B$. The collection of all of those arrows (or ordered pairs) determines a relation from the first set to the second set.

Definition A *binary relation R* from a set A to a set B is a nonempty subset of $A \times B$ (and any nonempty subset of $A \times B$ is a binary relation from A to B). A subset of $A \times A$ is called a binary relation *on A*.

(Perhaps you would feel more comfortable with this definition if the word "determines" were used in place of "is" in the first sentence. You should feel free to say "determines" if that helps in your thinking about binary relations.)

Some students may participate in several activities, some in none; also, some activities may attract no participants. The definition of a binary relation R tells us only that R is some subset of all the possible ordered pairs of the form (a, b) for $a \in A$ and $b \in B$.

Definition If R is a binary relation from A to B, we shall write $a \, R \, b$ if and only if the ordered pair (a, b) belongs to the set R. The *domain* of R is the set of all first entries in the set R of ordered pairs, and the *range* of R is the set of all second entries in the set R of ordered pairs. That is,

Domain of $R = \{a \in A \mid (a, b) \in R$ for at least one $b \in B\}$,
Range of $R = \{b \in B \mid (a, b) \in R$ for at least one $a \in A\}$.

A binary relation R from A to B can be represented by drawing two sets, labeled A and B, and an arrow from each element a in the domain of R to each element b in B such that $a \, R \, b$. For example, let $A = \{1, 2, 3, 4, 5\}$, and let $B = \{e, f, g, h, i, n, o, r, s, t, u, v, w\}$. Figure 2.1-1 pictures the relation R that associates each of the first three digits with each letter used in the spelling of the name of that digit.

Thus the domain of R is $\{1, 2, 3\}$, and the range is $\{e, h, n, o, r, t, w\}$. The ordered pairs that define R are

$(1, o)$	$(2, t)$	$(3, t)$
$(1, n)$	$(2, w)$	$(3, h)$
$(1, e)$	$(2, o)$	$(3, r)$
		$(3, e)$

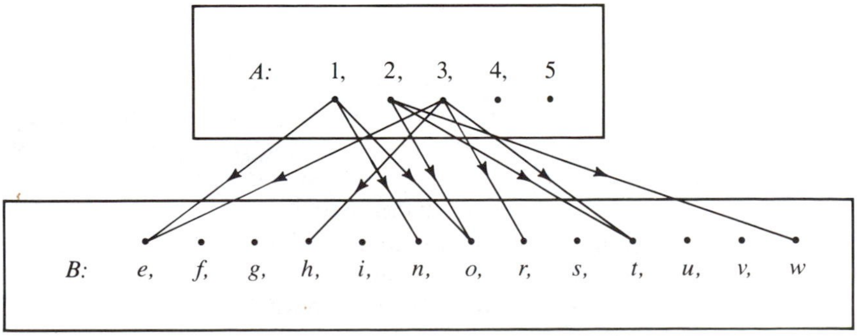

Figure 2.1-1

EXAMPLES   **1.** Let A denote the set of all residents of Michigan, and B the set of all doctors licensed to practice in Michigan. Let $a\ P\ b$, read "a is a patient of b," mean that person a was treated professionally at least once by doctor b.

2. Let P denote the set of all points on a plane, and T the set of all triangles on that plane. Let $a\ I\ b$ mean that "a is an interior point of b."

3. Let E denote the set of all entrants of a tennis tournament, and for all $a,\ b \in E$ let us write $a\ D\ b$ to mean that player a defeated player b in the tournament. In this example, D is a binary relation on E. ∎

Other examples of binary relations on a set are perpendicularity of lines in a plane, the relation "greater than" for real numbers, the relation "divides evenly into" for positive integers, and kinship among a set of people.

PROBLEM   How many different binary relations can be defined from a finite set A to a finite set B? ∎

Recalling that any subset of $A \times B$ is a binary relation from A to B, we can restate this question: "How many different subsets does $A \times B$ have?" Recall from Section 1.4 that $A \times B$ has $n(A)n(B)$ elements, so the question can be stated again as, "How many different subsets does a k-element set have?" where $k = n(A)n(B)$. This question was posed as Exercise 9 of Section 1.4; if your answer was 2^k, you were correct. Therefore the number of different binary relations from A to B is simply $2^{n(A)n(B)}$. If A has only three elements and B has only four, the number of possible binary relations from A to B is

$$2^{(3)(4)} = 2^{12} = 4096.$$

We wouldn't want to write a list of all these possibilities simply to count the number of relations from A to B.

Now that we know how many different binary relations can be defined, we see (even for small sets) that it might be useful to classify binary relations according to properties that occur frequently in relations of intrinsic interest. In doing so, we shall confine our attention to the special (but most frequent) case in which $B = A$.

Let A denote a set on which is defined a binary relation R. We define R to have the property listed in the left-hand column of Table 2.1-1 if and only if the corresponding condition in the right-hand column is satisfied for all $a,\ b,\ c$ in A.

Property	Condition: For all *a, b, c in A*
1. Reflexivity	*a R a*
2. Irreflexivity	*a R̸ a*, where *R̸* denotes "not related."
3. Symmetry	If *a R b*, then *b R a*.
4. Antisymmetry	If *a R b* and *b R a*, then *a = b*.
5. Asymmetry	If *a R b*, then *b R̸ a*.
6. Transitivity	If *a R b* and *b R c*, then *a R c*.

Table 2.1-1

Observe carefully that any of these properties is ascribed to a relation R on A only when the corresponding condition is satified for all choices of elements of A, including choices in which a, b, and c are not necessarily distinct. Also observe that most binary relations on A will have none of these properties because *every* subset of $A \times A$ determines a binary relation.

EXAMPLES
4. Parallelism of lines in the plane is symmetric. Depending on the way "parallel" is defined, it can be reflexive and transitive, or irreflexive and not transitive.

5. "Is the father of" is irreflexive and asymmetric, and not transitive.

6. In the natural numbers, the relation $<$ is irreflexive, asymmetric, and transitive; the relation $\le$ is reflexive, antisymmetric, and transitive. But the relation "is the square of" is neither reflexive, nor irreflexive, nor transitive. Observe that it is antisymmetric. Why? ∎

APPLICATION
In a basketball league assume that each team plays every other team an odd number of times in a season. Because ties cannot occur in basketball, this implies that at the end of the season, given any pair of teams, one will have won a majority of the games between those two teams. The majority winner is called the *dominant* team of that pair. What properties can be used to describe dominance?

Write $A \, d \, B$ to denote that team A dominates team B. Then $A \, d \, A$ is not sensible because no team is scheduled to play itself. Hence d is irreflexive. Also d satisfies "If $B \ne A$, then either $A \, d \, B$ or $B \, d \, A$ but not both." What about transitivity? If $A \, d \, B$ and $B \, d \, C$, can we conclude that $A \, d \, C$? The answer is no. This answer should not be surprising (because upsets do occur); it also makes the choice of a champion more difficult when transitivity fails to hold in a given season. How would you choose a champion from the results for the season as given in Table 2.1-2? First make a table of dominance pairs.

Winners \ Losers	A	B	C	D	E	F
A	0	1	3	2	1	1
B	2	0	1	1	3	1
C	0	2	0	3	1	1
D	1	2	0	0	2	2
E	2	0	2	1	0	2
F	2	2	2	1	1	0

Table 2.1-2   ∎

Exercises 2.1

1. Consider the following binary relations, defined on the set of all people now alive. For each of the relations determine which of the six properties it possesses and which it does not possess. If a relation does not possess a certain property, explain why.

   (a) "Is a sister of"
   (b) "Is a brother or sister of"
   (c) "Is a brother or sister of, or is the same person as"
   (d) "Is taller than or older than"
   (e) "Lives within 300 miles of"

2. "Asymmetry" of R means that $a\ R\ b$ and $b\ R\ a$ holds for no pair of elements a and b, but "antisymmetry" means that $a\ R\ b$ and $b\ R\ a$ holds if and only if $a = b$.

   (a) Describe how each of those two nonsymmetric properties is related to reflexivity or irreflexivity.
   (b) Can a relation be both asymmetric and antisymmetric? Explain your answer fully.

3. Define the binary relation R on the set N of natural numbers by writing

   $$a\ R\ b \quad \text{if and only if} \quad a \text{ is the square of } b.$$

   This relation was mentioned in Example 6.

   (a) Verify that R is antisymmetric.
   (b) Verify that R is not reflexive, not irreflexive, and not transitive.

4. Let A be the set of all points in a plane, and let p_0 be a fixed point in the plane. What properties in Table 2.1-1 does each of the following binary relations on A have, and what properties does it not have?

    (a) *a R b* if and only if *a* and *b* are equidistant from p_0.

    (b) *a R b* if and only if the distance from *b* to p_0 is greater than or equal to the distance from *a* to p_0.

5. Let $A = \{0\}$ and $B = \{0, 1, 2\}$. List all binary relations from A to B, stating the domain and range for each relation.

6. Let $A = \{0, 1, 2\}$. What properties listed in Table 2.1-1 does each of the following binary relations on A have? What properties does it not have? Justify your answers.

    (a) $R(1) = \{(0, 0), (1, 1), (2, 2), (1, 2), (2, 1)\}$.

    (b) $R(2) = \{(1, 1), (2, 2), (1, 2), (2, 1)\}$.

    (c) $R(3) = \{(0, 0), (1, 1), (2, 2), (0, 1), (1, 2)\}$.

    (d) $R(4) = \{(0, 0), (0, 1), (0, 2), (1, 1), (1, 2), (2, 2)\}$.

    (e) $R(5) = \{(0, 1), (0, 2), (1, 2)\}$.

    (f) $R(6) = \{(0, 0), (1, 1)\}$.

7. If *p* and *q* are statements, we define a binary relation R by writing

$$p \, R \, q \quad \text{if and only if} \quad p \text{ implies } q.$$

Is the relation R reflexive on the set of all statements? Is R symmetric? Is R transitive? Justify your answers.

8. (a) Make a table of dominance pairs for the final application of this section. Label the rows and columns from A through F; enter the symbol *d* in row X and column Y if and only if team X dominated team Y in that season.

    (b) Show that this dominance relation is not transitive.

    (c) Which team would you choose as champion for the season? Why?

9. Let $A = \{0, 1, 2, 3\}$. Then the set $R = \{(0, 1), (1, 2), (2, 1), (2, 3)\}$ is a subset of $A \times A$, and hence R is a binary relation on A.

    (a) Show that R is not transitive.

    (b) Which ordered pairs must be adjoined to R to have the enlarged set satisfy the transitivity condition? Explain your reasoning.

10. Let $A = N \times N$, where N is the set of natural numbers. Define a binary relation R on A by writing

$$(a, b) \, R \, (c, d) \quad \text{if and only if} \quad a + d = b + c.$$

Show that R is reflexive, symmetric, and transitive.

11. (a) By using an example show that a symmetric and transitive binary relation is not necessarily reflexive.

(b) What is wrong with the following argument? By the symmetric property, *a R b* and *b R a*. By transitivity, *a R b* and *b R a* imply *a R a*. Hence, if a relation is symmetric and transitive, then it is also reflexive.

2.2 Representations of Binary Relations

When *A* and *B* are finite sets, a binary relation *R* from *A* to *B* can be represented in forms that organize the information conveniently. One method is to list the members of *A* and those of *B* in parallel columns and draw an arrow from *a* to *b* whenever *a R b*. This procedure was described at the start of Section 2.1 but was abandoned as being too hard to unravel if *A* and *B* are not both small. We now consider a slight modification of that method for the case in which *B = A*; that is, when *R* is a binary relation on a single set.

Representation of a Binary Relation by a Directed Graph

Let *V* be a set of *n* distinct elements, which, for convenience, we denote by $\{1, 2, \ldots, n\}$. We can represent each element of *V* by a point on the plane, labeled by the numeral for that element of *V*. These *n* labeled points are called *vertices* of a *directed graph G*. For each ordered pair (i, j) of vertices, we draw a directed line segment from *i* to *j* if and only if *i R j*. If *i R i*, we draw a loop segment from *i* back to *i*. Each directed segment from vertex to vertex of *G* is called a *directed edge* of *G*, and the direction is indicated by an arrowhead along the directed edge (i, j). If *i R j* and *j R i*, then an arrowhead is placed in each direction along the segment between *i* and *j*.

> **Definition** A *directed graph G* consists of a nonempty set *V* whose elements are called the *vertices* of *G* together with a subset *E* of $V \times V$. If $a, b \in V$ and $(a, b) \in E$, then (a, b) is called a *directed edge* from *a* to *b*. An edge of the form (a, a) is called a *loop*.

Although the definition does not specify that *V* must be finite, only that case will be considered in this treatment of graph theory.

EXAMPLE 1 If you made a table of dominance pairs for the application at the end of Section 2.1, you should have obtained a table similar to Table 2.2-1.

	A	B	C	D	E	F
A	—	—	d	d	—	—
B	d	—	—	—	d	—
C	—	d	—	d	—	—
D	—	d	—	—	d	d
E	d	—	d	—	—	d
F	d	d	d	—	—	—

Table 2.2-1

Thus A dominates C and D, B dominates A and E, and so on. To represent this dominance relation d by a directed graph G, we select six points that we label A through F (instead of switching notation to 1 through 6). It usually works well to choose the vertices of G as though they were spread evenly along a circle (see Figure 2.2-1). Then the directed edges are added in conformity with the table of dominance, Table 2.2-1.

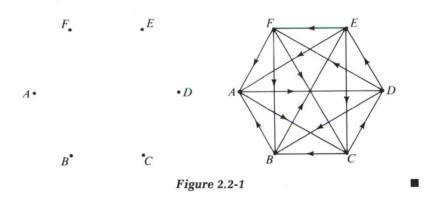

Figure 2.2-1 ■

Observe that the graph of a dominance relation D on a set S will have

1. no loops, because $j \, D \, j$ holds for no $j \in S$;
2. one edge, directed in only one way, between each pair of distinct vertices, because for each distinct pair $(i, j) \in S \times S$ either $i \, D \, j$ or $j \, D \, i$ but not both.

Thus the properties of G accurately reflect the properties of the binary relation that G represents.

Of course, if R is a symmetric relation on a set S, then the graph of R will have the property that each edge of G will be directed in both directions, because $i \, R \, j$ implies $j \, R \, i$ by symmetry. When each edge of a directed graph G is directed in both directions, there is no need to put arrowheads on the edges, so the arrowheads are omitted, and each edge is undirected. We also omit the word "directed" and simply speak of a graph.

> **Definition**   A *graph* G consists of a nonempty set V (the *vertices* of G) and a subset E (the *edges* of G) of $V \times V$ such that $(i, j) \in E$ if and only if $(j, i) \in E$.

Up to this point we have started with a given binary relation on a set and described how we can obtain its directed graph. Conversely, given any directed graph G, there is a set S and a relation R such that G is the graph of R. To see this, we choose S to be the set V of vertices of G, and for any pair $(i, j) \in V \times V$ we write

$$i \, R \, j \quad \text{if and only if} \quad (i, j) \text{ is an edge of } G.$$

Thus the study of directed graphs can be regarded as a study of binary relations that can be defined on the set of vertices of the graph. As we shall see, graphs can be used to represent many different situations. For instance, to a psychologist a directed graph like that of Figure 2.2-1 might suggest the lines of influence that could exist among six persons serving together on the town council. After having listened to one another through many long and sometimes heated sessions, persons tend to form relationships of respect (or otherwise) that affect their views. Most of us have friends whose opinions influence us and acquaintances whose views we ignore. Suppose we look at a different example.

EXAMPLE 2   **Influence within a Committee**

Consider a committee C of five persons, $C = \{a, b, c, d, e\}$. We represent the web of influence within C by drawing a directed graph with C as its set of vertices and an edge directed from x to y if and only if x directly influences y, as shown in Figure 2.2-2 for instance.

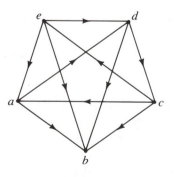

Figure 2.2-2   ■

Note that for each pair (x, y) of committee members either x influences y or y influences x. That condition expresses the idea that there is at least one-way influence between each pair of persons. Under those conditions the following stronger result is known to be valid.

Theorem 2.2-1 In any committee in which, for each pair (x, y) of distinct members, either x influences y or y influences x, there is at least one member who influences every other member, either directly or indirectly through one intermediary.

For example, using Figure 2.2-2 we can observe that c influences $a, b,$ and e directly, and c influences d indirectly through either a or e. In Section 2.3 we shall demonstrate this same result in a different way, after having introduced a second method of representing binary relations and, therefore, of representing directed graphs.

Representation of a Binary Relation by a 0-1 Matrix

As before, let S denote a set of n distinct elements, $S = \{1, 2, \ldots, n\}$, and let R denote a binary relation on S. We form a square array with n rows (horizontal) numbered $1, 2, \ldots, n$ and n columns (vertical) numbered $1, 2, \ldots, n$.

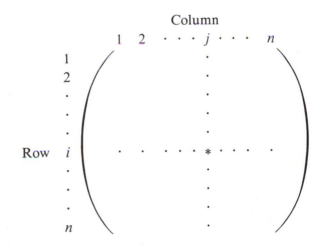

At the intersection of row i and column j we write 1 if and only if $i\,R\,j$. Otherwise we write 0 in that position. And we use this rule to enter either 0 or 1 in each of the n^2 positions of the table. This table is called a *square 0-1 matrix of size n*. For the binary relation R represented by the directed graph in Figure 2.2-1, the 0-1 matrix M that represents R can be obtained easily from Table 2.2-1 by replacing each d entry by 1 and each blank entry by 0:

$$M = \begin{pmatrix} 0 & 0 & 1 & 1 & 0 & 0 \\ 1 & 0 & 0 & 0 & 1 & 0 \\ 0 & 1 & 0 & 1 & 0 & 0 \\ 0 & 1 & 0 & 0 & 1 & 1 \\ 1 & 0 & 1 & 0 & 0 & 1 \\ 1 & 1 & 1 & 0 & 0 & 0 \end{pmatrix}.$$

Similarly, for the influence relation represented by the graph in Figure 2.2-2, we obtain the entries of the corresponding matrix J having five rows and five columns:

$$J = \begin{array}{c} \\ a \\ b \\ c \\ d \\ e \end{array} \begin{array}{ccccc} a & b & c & d & e \\ \begin{pmatrix} 0 & 1 & 0 & 1 & 0 \\ 0 & 0 & 0 & 0 & 0 \\ 1 & 1 & 0 & 0 & 1 \\ 0 & 1 & 1 & 0 & 0 \\ 1 & 1 & 0 & 1 & 0 \end{pmatrix}. \end{array}$$

This matrix represents the pattern of direct (one-step) influence within the committee. For example, person a directly influences b and d but no one else; hence in the first row (labeled a) the numeral 1 is entered in the columns labeled b and d, and 0 is entered in every other column of row 1. Because b does not influence any committee member, 0 is entered in each column of row 2, and so on.

Next let's examine how two-step influence (through one intermediary) can occur in this example. Specifically, person a influences only b and d in one step; hence a can influence c in two steps only if either b or d can influence c directly. We see that d directly influences c, but b does not. Hence, a can exert two-step influence on c in one and only one way — through d. Observe that a cannot exert two-step influence on e because neither b nor d exerts one-step influence on e.

Again referring to the graph in Figure 2.2-2, we observe that c influences b, a, and e in one step, and c influences d in two steps along two paths — through a and through e. Hence c influences everyone in either one or two steps.

It is possible, and often easier, to obtain such information numerically from the matrix J rather than from the graph that J represents. We shall do that in the next section after we learn how to calculate with matrices. Notice, however, that once again we are using the numerals 0 and 1 to represent information that initially appears not to be numerical — the inability or ability of person x to influence person y directly. Notice also that the numerical entries of a 0-1 matrix can be entered easily into a computer. Because computers can be programmed to carry out matrix calculations, they can be used to analyze the information implicit in binary relations (or directed graphs).

Although a 0-1 matrix is a simple way to represent a graph within a computer, it is highly inefficient because it contains so much redundant data. If we know which entries of a 0-1 matrix are 1, we know that every other entry is 0. If you study computer science, you will learn about other data structures that are more efficient than 0-1 matrices for representing graphs in computers.

Exercises 2.2

1. Let *G* denote the following directed graph.

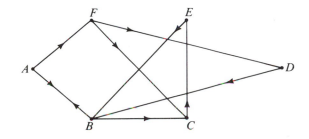

(a) The directed graph *G* represents a binary relation *R* on the set of vertices of *G*. Use set notation to describe *R*.
(b) Is *R* a transitive relation? Explain.
(c) Write the 0-1 matrix that represents *R*, labeling the rows and columns of the matrix with the letters used for the vertices of *G*.

2. Do Exercise 1 with the directed graph replaced by the following graph.

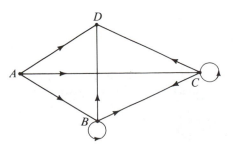

3. Let *R* be a binary relation on the set *S* = {*A*, *B*, *C*, *D*}, as represented by the 0-1 matrix *M*:

$$M = \begin{array}{c} \\ A \\ B \\ C \\ D \end{array} \begin{array}{cccc} A & B & C & D \\ \left(\begin{array}{cccc} 0 & 0 & 0 & 1 \\ 1 & 0 & 0 & 1 \\ 0 & 1 & 0 & 1 \\ 0 & 1 & 1 & 0 \end{array}\right). \end{array}$$

(a) Use set notation to describe R.

(b) Draw a directed graph that represents R.

4. Do Exercise 3 with S replaced by $S = \{A, B, C, D, E, F\}$ and the 0-1 matrix replaced by

$$M = \begin{array}{c} \\ A \\ B \\ C \\ D \\ E \\ F \end{array} \begin{array}{cccccc} A & B & C & D & E & F \\ \left(\begin{array}{cccccc} 0 & 1 & 0 & 1 & 0 & 1 \\ 0 & 0 & 0 & 1 & 1 & 1 \\ 0 & 0 & 0 & 1 & 0 & 1 \\ 1 & 1 & 0 & 0 & 0 & 0 \\ 1 & 0 & 1 & 0 & 0 & 0 \\ 0 & 0 & 0 & 0 & 1 & 0 \end{array}\right). \end{array}$$

5. Draw the associated directed graph and write the associated 0-1 matrix for each of the binary relations given in Exercise 6 of Section 2.1.

6. Let $S = \{0, 1, 2\}$ and let $\mathscr{F}$ denote the family of all subsets of S. We define a binary relation R on $\mathscr{F}$ by writing

$$A\,R\,B \quad \text{if and only if} \quad A \subseteq B \text{ and } A \neq B.$$

(a) List the elements of $\mathscr{F}$ and draw a directed graph representing R.

(b) Write the 0-1 matrix that represents R.

7. Let $A = \{2, 3, 5, 8, 10, 12, 36, 40\}$. Let R be the binary relation defined on A by

$$a\,R\,b \quad \text{if and only if} \quad a \text{ divides } b \text{ and } a \neq b.$$

(a) Draw a directed graph that represents R.

(b) Write the 0-1 matrix that represents R.

8. Let R be a binary relation on a set A.

(a) How are the properties of reflexivity and symmetry exhibited in the graph that represents R?

(b) How are these two properties exhibited in the 0-1 matrix that represents R?

9. A commuter airline serves seven cities. Denote the names of the cities by $A(1)$, $A(2)$, . . . , $A(7)$. There are direct flights between only certain pairs of cities. The following 0-1 matrix gives the direct flights, where a 1 in the ith row and jth column means there is a direct flight from city $A(i)$ to city $A(j)$, whereas a 0 indicates there is no direct flight.

$$M = \begin{array}{c} \\ A(1) \\ A(2) \\ A(3) \\ A(4) \\ A(5) \\ A(6) \\ A(7) \end{array} \begin{pmatrix} A(1) & A(2) & A(3) & A(4) & A(5) & A(6) & A(7) \\ 0 & 1 & 0 & 1 & 0 & 0 & 0 \\ 0 & 0 & 1 & 0 & 1 & 0 & 0 \\ 1 & 0 & 0 & 0 & 0 & 1 & 1 \\ 0 & 0 & 1 & 0 & 0 & 1 & 0 \\ 0 & 0 & 1 & 0 & 0 & 1 & 0 \\ 0 & 0 & 0 & 0 & 0 & 0 & 1 \\ 0 & 0 & 0 & 0 & 1 & 0 & 0 \end{pmatrix}.$$

(a) Draw the graph associated with the given 0-1 matrix M.
(b) Suppose you live in $A(1)$. In how many ways can you fly to city $A(3)$ with only one intermediate city? Assume that your arrival occurs before the departure of the connecting flight.
(c) If you live in $A(1)$, how many ways are there to fly to city $A(7)$ with three or fewer intermediate cities? Again, assume no conflicts in arrival and departure times.

10. Consider a set of five persons $P(1)$, $P(2)$, $P(3)$, $P(4)$, $P(5)$ in a volunteer fire department. We write $P(i) \, R \, P(j)$ if and only if there is a one-way communication link from $P(i)$ to $P(j)$. The directed graph G of the binary relation R is given by the sketch.

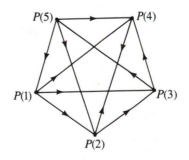

(a) Write the 0-1 matrix M that represents G.
(b) Show that for every pair $(P(i), P(j))$ of the five persons there is a one-way communication link from $P(i)$ to $P(j)$ or from $P(j)$ to $P(i)$. What can you conclude from this? Why?

(c) Which persons, if any, can communicate with every other person either directly or indirectly through only one intermediate person? Justify your answer.

11. Consider the network of one-way streets given in the figure, where the arrows indicate the direction of traffic.

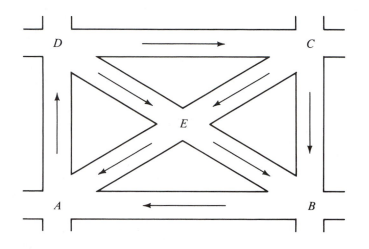

(a) Represent the network by a directed graph G whose vertices are the five labeled intersections and where an edge directed from X to Y is drawn if and only if one can travel from X to Y in the correct direction without crossing an intersection.

(b) Write the 0-1 matrix associated with the graph G.

(c) From which intersections, if any, can one travel to every other intersection either directly or indirectly through only one intermediate intersection?

12. Four people $P(1)$, $P(2)$, $P(3)$, $P(4)$ play in a round robin tennis match (i.e., every two people play each other and no ties are allowed). The results of the match are as follows:

$P(1)$ beats $P(4)$

$P(2)$ beats $P(1)$ and $P(4)$

$P(3)$ beats $P(1)$ and $P(2)$

$P(4)$ beats $P(3)$

Write $P(i)\, d\, P(j)$ if and only if $P(i)$ beats $P(j)$.

(a) Draw a graph G that represents the binary relation d.

(b) Write the 0-1 matrix for d.

(c) Suppose we say that $P(i)$ has a quasi-victory over $P(j)$ if $P(i)$ beats a player who has in turn beaten $P(j)$. Note that $P(i)$ may beat $P(j)$ and at the same time have a quasi-victory over $P(j)$. How many quasi-victories does each player have? Show your reasoning.

(d) If you rank the players according to the sum of the number of victories and the number of quasi-victories of each player, what would the ranking be?

(e) Show that players $P(2)$ and $P(3)$ have been defeated only by players who themselves have lost to players defeated by $P(2)$ and $P(3)$.

2.3   Elements of Matrix Calculations

To use 0-1 matrices in our study of either directed graphs or binary relations, we must first learn the rudiments of matrix arithmetic. Our discussion will be limited to square matrices of real numbers.

Definition   Let n be a positive integer. An *n-by-n matrix A* of real numbers is a square array of n^2 real numbers arranged in n horizontal rows and n vertical columns. The rows are numbered from top to bottom and the columns from left to right, each with the numerals $1, 2, 3, \ldots, n$. For each ordered pair (i, j), where i and j range independently from 1 to n, the number in row i and column j of A is denoted by $a(i, j)$, and we write

$$A \;=\; \text{row } i \begin{pmatrix} & & \vdots & & \\ \cdots & \cdots & a(i, j) & \cdots & \cdots \\ & & \vdots & & \end{pmatrix}.$$

EXAMPLE 1   Let

$$A = \begin{pmatrix} 5 & 1.07 & -\pi \\ 0 & -2 & 1 \\ 4 & 3/2 & 0 \end{pmatrix}.$$

Then A is a 3-by-3 matrix. The entry $a(3, 1)$ is 4, and the entry $a(1, 3)$ is $-\pi$. Also $a(2, 1) = 0 = a(3, 3)$, and so on.   ■

Before describing matrix arithmetic, we define equality of two n-by-n matrices, A and B. We denote the (i, j) entry of A by $a(i, j)$ and the (i, j) entry of B by $b(i, j)$ for $i = 1, 2, \ldots, n$ and $j = 1, 2, \ldots, n$.

Definition   Two n-by-n matrices are *equal* if and only if entries in corresponding positions are equal for every position. Symbolically, $A = B$ if and only if $a(i, j) = b(i, j)$ for all i and j from 1 to n, inclusive.

There are three basic arithmetic operations that can be performed on square matrices:

1. The numerical multiple cA of a number c and a matrix A,
2. The sum $A + B$ of two n-by-n matrices A and B,
3. The matrix product AB of two n-by-n matrices.

Both 1 and 2 are quite easy to perform, but 3 might be unfamiliar to you and therefore require your close attention. To define these operations, we let c be any real number and let A and B be n-by-n matrices for any fixed positive integer n.

Definitions

1. The (i, j) entry of cA is $ca(i, j)$. To multiply A by a number c, multiply each entry of A by c.
2. The (i, j) entry of $A + B$ is $a(i, j) + b(i, j)$. To add two n-by-n matrices, add the two corresponding entries position by position.
3. The (i, j) entry of AB is

$$a(i, 1)b(1, j) + a(i, 2)b(2, j) + \cdots + a(i, n)b(n, j).$$

To multiply two n-by-n matrices, calculate the (i, j) entry of AB as the sum of all products $a(i, k)b(k, j)$ as k varies from 1 to n.

Fortunately, the product of two matrices is much easier to calculate in practice than you might expect from the formal definition, as we shall now demonstrate for the case where $n = 3$.

EXAMPLE 2 Let

$$c = -3, \quad A = \begin{pmatrix} 1 & -2 & 5 \\ -3 & 4 & 0 \\ 1 & -6 & 7 \end{pmatrix}, \quad B = \begin{pmatrix} 5 & 3 & -4 \\ 2 & 2 & 0 \\ 0 & -1 & 1 \end{pmatrix}.$$

Then

$$-3A = \begin{pmatrix} -3 & 6 & -15 \\ 9 & -12 & 0 \\ -3 & 18 & -21 \end{pmatrix};$$

$$A + B = \begin{pmatrix} 1+5 & -2+3 & 5-4 \\ -3+2 & 4+2 & 0+0 \\ 1+0 & -6-1 & 7+1 \end{pmatrix} = \begin{pmatrix} 6 & 1 & 1 \\ -1 & 6 & 0 \\ 1 & -7 & 8 \end{pmatrix}.$$

To illustrate the process of matrix multiplication, let us first calculate the $(3, 2)$ entry of AB. In the product definition let $i = 3$ and $j = 2$; compute

$$a(3, 1)b(1, 2) + a(3, 2)b(2, 2) + a(3, 3)b(3, 2)$$
$$= (1)(3) + (-6)(2) + (7)(-1) = -16.$$

Observe that i determines the row and j the column of the entry of AB currently being calculated. Now look at row i of the left-hand matrix A of the product AB, and look at column j of the right-hand matrix B of AB. Specifically, look at row 3 of A and column 2 of B.

$$\rightarrow \quad \begin{pmatrix} \cdot & \cdot & \cdot \\ \cdot & \cdot & \cdot \\ 1 & -6 & 7 \end{pmatrix} \quad \begin{pmatrix} \cdot & 3 & \cdot \\ \cdot & 2 & \cdot \\ \cdot & -1 & \cdot \end{pmatrix}$$

Consider the chosen row and the chosen column as two ordered triples of numbers; multiply the two first numbers of each triple, then multiply the two second numbers, and finally multiply the two last numbers. The sum of those three products is the $(3, 2)$ entry of AB, namely -16.

Thus, the $(2, 1)$ entry of AB is -7, calculated as shown.

$$\downarrow$$

$$\rightarrow \quad \begin{pmatrix} \cdot & \cdot & \cdot \\ -3 & 4 & 0 \\ \cdot & \cdot & \cdot \end{pmatrix} \quad \begin{pmatrix} 5 & \cdot & \cdot \\ 2 & \cdot & \cdot \\ 0 & \cdot & \cdot \end{pmatrix};$$

$$(-3)(5) + (4)(2) + (0)(0) = -7.$$

Verify that the remaining entries are as follows:

$$AB = \begin{pmatrix} 1 & -6 & 1 \\ -7 & -1 & 12 \\ -7 & -16 & 3 \end{pmatrix}.$$

The product AB of two n-by-n matrices is calculated in the same way, except that each row and each column has n entries instead of three. ■

The (i, j) entry of the product AB can be written compactly by using summation notation, which we now explain and illustrate. The Greek letter sigma, written Σ, corresponds to the Latin letter S and is used as an instruction to sum all of the numbers that can be formed by assigning consecutive integer values to a variable that appears in the formula following Σ. The name of that variable and its starting value are written directly under Σ, and the final value of the variable is written directly above Σ. For example,

$$\sum_{k=1}^{n} k^2 \quad \text{means} \quad 1^2 + 2^2 + 3^2 + \cdots + n^2.$$

Study the following illustrations carefully:

$$\sum_{k=3}^{9} k(k+1) = (3)(4) + (4)(5) + (5)(6) + \cdots + (9)(10) = 322;$$

$$\sum_{k=1}^{m} (3k + i) = [3(1) + i] + [3(2) + i] + [3(3) + i] + \cdots + [3(m) + i];$$

$$\sum_{i=1}^{m} (3k + i) = [3k + 1] + [3k + 2] + [3k + 3] + \cdots + [3k + m];$$

$$\sum_{k=1}^{n} a(i, k) = a(i, 1) + a(i, 2) + a(i, 3) + \cdots + a(i, n).$$

Note that in the last three summations the expression to be summed contains two symbols, i and k, that could be regarded as variables; in each case the variable of summation is the letter written beneath Σ, and the other symbol is regarded as a number that is independent of the summation variable. The final equation can be interpreted as the sum of the entries in row i of an n-by-n matrix A.

Now we are ready to use the summation notation to express the (i, j) entry of the product AB of two n-by-n matrices, where we regard i and j as arbitrary but *fixed* integers from 1 to n, inclusive. By definition the (i, j) entry is

$$a(i, 1)b(1, j) + a(i, 2)b(2, j) + \cdots + a(i, n)b(n, j).$$

Note that each term in this sum has the form $a(i, *)b(*, j)$; that is, the second index in $a(i, *)$ is equal to the first index in $b(*, j)$. Furthermore, the values that are to be assigned to $*$ are successive integers from 1 through n. We can denote the variable index $*$ by any letter that is not otherwise used in the formula to be summed; in this case k is a suitable choice. Then each term to be summed has the form $a(i, k)b(k, j)$ for some k from 1 through n. Thus the (i, j) entry of AB can be written

$$\sum_{k=1}^{n} a(i, k)b(k, j).$$

EXAMPLE 3   **The Two-Step Influence Matrix**

We return to the direct (or one-step) influence matrix J of the graph in Figure 2.2-2.

$$
J = \begin{array}{c} \\ a \\ b \\ c \\ d \\ e \end{array}
\begin{array}{c} \begin{array}{ccccc} a & b & c & d & e \end{array} \\
\left(\begin{array}{ccccc}
0 & 1 & 0 & 1 & 0 \\
0 & 0 & 0 & 0 & 0 \\
1 & 1 & 0 & 0 & 1 \\
0 & 1 & 1 & 0 & 0 \\
1 & 1 & 0 & 1 & 0
\end{array} \right). \end{array}
$$

By the result stated in the paragraphs just below Figure 2.2-2, one or more of the committee members has the capacity to influence all the committee members either directly (in one step) or through an in⸱ermediary (in two steps). Now let's attempt to analyze the two-step influence process in matrix form.

Member a can influence members b and d directly. We ask whether a has any two-step influence either with c or with e. For a to have two-step influence with c, either b or d must have direct influence with c, which occurs if 1 appears in b's row and c's column, or if 1 appears in d's row and c's column. The latter does occur. Similarly, whether or not a has two-step influence with e depends on whether there is a 1 in b's row and e's column, or else in d's row and e's column. Because both of those entries are 0, we conclude that a does not have any two-step influence with e.

To see what is happening numerically, we write the entries of a's row, and then directly below we write the entries for c's column.

$$
\begin{array}{lccccc}
a\text{'s row:} & 0 & 1 & 0 & 1 & 0 \\
c\text{'s column:} & 0 & 0 & 0 & 1 & 0
\end{array}
$$

If we form the sum of the term-by-term products of this pair of quintuples we obtain 1, indicating that a has one path of two-step influence with c (through d since 1 is the product of the digits in the fourth positions). But a has no two-step path to e because a can directly influence only b or d, and neither b nor d can directly influence e. This is revealed numerically by the fact that 0 is the sum of the term-by-term products of the following pair of quintuples.

$$
\begin{array}{lccccc}
a\text{'s row:} & 0 & 1 & 0 & 1 & 0 \\
e\text{'s column:} & 0 & 0 & 1 & 0 & 0
\end{array}
$$

These calculations simply mean that in the matrix product J^2 the number 1 is in row a and column c, and 0 is in row a and column e. You may verify by matrix multiplication that

$$
J^2 = \begin{pmatrix} 0 & 1 & 0 & 1 & 0 \\ 0 & 0 & 0 & 0 & 0 \\ 1 & 1 & 0 & 0 & 1 \\ 0 & 1 & 1 & 0 & 0 \\ 1 & 1 & 0 & 1 & 0 \end{pmatrix} \begin{pmatrix} 0 & 1 & 0 & 1 & 0 \\ 0 & 0 & 0 & 0 & 0 \\ 1 & 1 & 0 & 0 & 1 \\ 0 & 1 & 1 & 0 & 0 \\ 1 & 1 & 0 & 1 & 0 \end{pmatrix} = \begin{pmatrix} 0 & 1 & 1 & 0 & 0 \\ 0 & 0 & 0 & 0 & 0 \\ 1 & 2 & 0 & 2 & 0 \\ 1 & 1 & 0 & 0 & 1 \\ 0 & 2 & 1 & 1 & 0 \end{pmatrix}.
$$

To see the meaning of the entries of J^2, recall that the 1 in the $(1, 3)$ position of J^2 indicates that there is one two-step path from a to c. How does the entry 2 in the $(3, 2)$ position occur, and what does it mean? It is the result of summing the term-by-term products of row 3 of J with column 2 of J.

$$
\begin{array}{lccccc}
\text{Row 3:} & 1 & 1 & 0 & 0 & 1 \\
\text{Col 2:} & 1 & 0 & 1 & 1 & 1 \\
\text{Products:} & 1 & 0 & 0 & 0 & 1 \\
\text{Sum of products:} & 2
\end{array}
$$

Observe that c has direct influence on a, b, e, whereas a has direct influence on b, and e has direct influence on b. The c to b path is counted already as a one-step influence because the (3, 2) entry of J is 1. The number of one-step paths of influence from one member to another is given for all pairs of members by the entries of J. Similarly, the number of two-step paths of influence between each pair of members is given by J^2. Thus the entry 2 in the (3, 2) position of J^2 means that there are two two-step paths from c to b. Hence, the matrix sum of J and J^2 will indicate the number of one-step or two-step influence paths from each member to each other member. We have

$$J + J^2 = \begin{pmatrix} 0 & 1 & 0 & 1 & 0 \\ 0 & 0 & 0 & 0 & 0 \\ 1 & 1 & 0 & 0 & 1 \\ 0 & 1 & 1 & 0 & 0 \\ 1 & 1 & 0 & 1 & 0 \end{pmatrix} + \begin{pmatrix} 0 & 1 & 1 & 0 & 0 \\ 0 & 0 & 0 & 0 & 0 \\ 1 & 2 & 0 & 2 & 0 \\ 1 & 1 & 0 & 0 & 1 \\ 0 & 2 & 1 & 1 & 0 \end{pmatrix}$$

$$= \begin{pmatrix} 0 & 2 & 1 & 1 & 0 \\ 0 & 0 & 0 & 0 & 0 \\ 2 & 3 & 0 & 2 & 1 \\ 1 & 2 & 1 & 0 & 1 \\ 1 & 3 & 1 & 2 & 0 \end{pmatrix}.$$

Notice that members c, d, and e can each exert one- or two-step influence on every other member. Of course, the result stated without proof in Theorem 2.2-1 guarantees only that there is at least one such influential member. ∎

Exercises 2.3

1. Let the matrices A and B be defined by the equations

$$A = \begin{pmatrix} 2 & 0 & 1 \\ -1 & 3 & 0 \\ 1 & 1 & 2 \end{pmatrix}, \qquad B = \begin{pmatrix} 2 & -3 & 1 \\ 1 & 1 & 0 \\ 0 & 2 & -1 \end{pmatrix}.$$

Calculate each of the following.

(a) $2A$
(b) $A + B$
(c) AB
(d) BA
(e) A^2

Note that the operation of multiplication of matrices is not a commutative operation.

2. Express the following sums without using summation notation.

(a) $\displaystyle\sum_{k=1}^{n} (2k+1)$

(b) $\displaystyle\sum_{k=1}^{n-1} k(k+2)^2$

(c) $\displaystyle\sum_{k=1}^{4} a(3, k)$

(d) $\displaystyle\sum_{k=1}^{n} a(i, k)a(k, j)$

3. Express the following sums by using summation notation.

(a) $1^3 + 2^3 + 3^3 + \cdots + (n+1)^3$

(b) $(1)(3)^2 + (2)(4)^2 + (3)(5)^2 + \cdots + (n)(n+2)^2$

(c) $(1)(3)^2 + (2)(4)^2 + (3)(5)^2 + \cdots + (n-2)(n)^2$

(d) $a(1, i)b(j, 1) + a(2, i)b(j, 2) + \cdots + a(5, i)b(j, 5)$

4. By tracing all the one-step and two-step paths from c to a, b, d, e on the graph in Figure 2.2-2, verify the numbers in the third row of the matrix $J + J^2$, calculated in the final paragraph of this section.

5. A small company has seven officers: a president (P), a vice-president (F) for financial affairs, a vice-president (D) for development, a comptroller (C), a director (L) of labor relations, a director (R) of public relations, and a director (S) of sales. If X and Y are officers, we write

$$X \, d \, Y \quad \text{if and only if} \quad X \text{ may request action by } Y.$$

The 0-1 matrix representing d is given by

$$
M = \begin{array}{c} \\ P \\ F \\ D \\ C \\ L \\ R \\ S \end{array}
\begin{array}{c}
\begin{array}{ccccccc} P & F & D & C & L & R & S \end{array} \\
\left(\begin{array}{ccccccc}
0 & 1 & 1 & 0 & 0 & 0 & 0 \\
0 & 0 & 1 & 1 & 1 & 0 & 1 \\
0 & 1 & 0 & 1 & 0 & 1 & 1 \\
0 & 0 & 0 & 0 & 0 & 0 & 0 \\
0 & 0 & 0 & 1 & 0 & 1 & 1 \\
0 & 0 & 0 & 1 & 0 & 0 & 0 \\
0 & 0 & 0 & 1 & 0 & 1 & 0
\end{array}\right).
\end{array}
$$

(a) Draw a directed graph representing d.

(b) Calculate the matrix $M + M^2$ and use it to determine the number of ways the vice-president for development can request action, either directly or through only one intermediate officer, by each of the officers except himself and the president.

6. (a) Calculate M^2 and $M + M^2$ for the matrix M given in Exercise 9 of Section 2.2.
   (b) By tracing all direct flights and all flights requiring exactly one intermediate city from $A(1)$ to $A(2)$, $A(3)$, $A(4)$, $A(5)$, $A(6)$, $A(7)$, verify the numbers in the first row of $M + M^2$.
   (c) Determine which city or cities you should choose to live in if you want to be able to fly to each of the other six cities with no more than one intermediate city. Explain your reasoning.
   (d) Determine which city or cities you should choose to live in if you want people in the other six cities to be able to fly to your home city with no more than one connecting flight. Explain.

7. (a) Calculate M^2 and $M + M^2$ for the matrix M of Exercise 10 in Section 2.2.
   (b) Determine which persons in the fire department can receive communications from each of the other members either directly or indirectly through exactly one intermediate person. Explain your reasoning.
   (c) Determine which member of the fire department has the greatest number of ways of sending messages either directly or indirectly through only one intermediate person. Show your calculations.
   (d) Do part c with the word "sending" replaced by "receiving." Show your calculations.

8. The president of Anon College is trying to convince the Loaded Foundation to endow three new positions in the Anon Department of Mathematics. There are seven members on the Loaded board, and a large grant requires the affirmative vote of all members. The web of influence within the board is reported to be as indicated in the following graph:

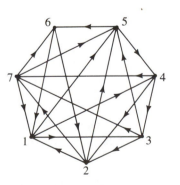

(a) Does this graph satisfy the hypothesis of Theorem 2.2-1?
(b) The president is permitted to present the proposal directly to any member of the board but only to that member. With which board member should the president choose to meet? Show your calculations and reasoning.
(c) If the board member referred to in **b** is unexpectedly absent, is there any other good choice? Explain.

9. Let A, B, C, Z be matrices defined by

$$A = \begin{pmatrix} 0 & 0 & 0 \\ 1 & 0 & 0 \\ 1 & 1 & 0 \end{pmatrix}, \qquad B = \begin{pmatrix} 1 & 2 & 0 \\ 1 & 0 & -2 \\ 0 & 1 & 1 \end{pmatrix},$$

$$C = \begin{pmatrix} 2 & -2 & -2 \\ -1 & 1 & 1 \\ 1 & -1 & -1 \end{pmatrix}, \qquad Z = \begin{pmatrix} 0 & 0 & 0 \\ 0 & 0 & 0 \\ 0 & 0 & 0 \end{pmatrix}.$$

The matrix Z is called the 3-by-3 *zero matrix*. From the definition of addition of matrices it is clear that $X + Z = X = Z + X$ for every 3-by-3 matrix X. Hence, in the arithmetic of 3-by-3 matrices, Z plays the same role as the number 0 plays in the arithmetic of real numbers.

(a) Show that $XZ = Z = ZX$ for every 3-by-3 matrix X.
(b) Calculate A^2 and $A^3 = AA^2$. If your work is correct, you will conclude that $A^2 \neq Z$ but $A^3 = Z$. Note that this phenomenon does not occur in the arithmetic of real numbers, for if a is a real number, then $a^3 = 0$ implies that $a = a^2 = 0$. (A matrix A for which $A^{n-1} \neq Z$ but $A^n = Z$ is said to be *nilpotent of index n*.)
(c) Show that $BC = Z \neq CB$. Note that this phenomenon does not occur in the arithmetic of real numbers, because the product of nonzero real numbers is never zero. (The matrices B and C are said to be *zero-divisors* because each is nonzero, but the product of each with a nonzero matrix is the zero matrix.)

10. Let A and I be matrices defined by

$$A = \begin{pmatrix} 1 & 0 & 0 \\ 0 & 1 & 0 \\ 0 & 0 & 0 \end{pmatrix}, \qquad I = \begin{pmatrix} 1 & 0 & 0 \\ 0 & 1 & 0 \\ 0 & 0 & 1 \end{pmatrix}.$$

(a) Show that $XI = X = IX$ for every 3-by-3 matrix X. The matrix I thus plays the same role in the arithmetic of 3-by-3 matrices as the number 1 plays in the arithmetic of real numbers. The matrix I is called the 3-by-3 *identity matrix*.
(b) Verify that $A^2 = A$. A matrix having this property is said to be *idempotent*. Note that in the arithmetic of real numbers $a^2 = a$

implies that $a = 1$ or $a = 0$. The matrix A in this exercise satisfies $A^2 = A$, and yet $A \neq I$ and $A \neq Z$. (For the definition of the matrix Z see Exercise 9.)

(c) In what other setting did the property of idempotence occur previously in this book?

11. Let J be the influence matrix given in Example 3 of this section. Give an interpretation for the entries of

(a) the matrix J^3,
(b) the matrix $J + J^2 + J^3$.

12. Let R be a binary relation defined on a set of four individuals a, b, c, d as follows:

$$x \, R \, y \quad \text{if and only if} \quad x \text{ can send a message to } y.$$

A directed graph that represents R is

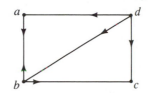

(a) Determine the matrix M that represents R.
(b) Calculate M^3, and use the graph to interpret the entry in the fourth row and third column.
(c) Calculate the matrix $M + M^2 + M^3$, and use the graph to interpret the entry in the fourth row and third column.

13. Let $A = (a(i, j))$ be a 0-1 matrix that represents a directed graph G having n vertices, and let $C = (c(i, j))$, where

$$c(i, j) = \sum_{k=1}^{n} a(i, k)a(k, j).$$

Use the fact that each number $a(r, s)$ is 0 or 1 to explain why $c(i, j)$ is the number of directed two-step paths from vertex i to vertex j in the graph G.

2.4 Equivalence Relations and Partitions

Equivalence is one of the most common binary relations. Indeed, there are many different senses of equivalence that could be applied sensibly to a given set. For example, consider the student body of College A. The registrar is charged with keeping the complete academic record of each student, and no two persons are regarded as identical. However, if special housing or class attendance regulations apply only to first-year students, then the administration or the faculty regards all first-year students as equivalent (in some particular sense) and all other students as equivalent (but different from the first-year students). The laws of a certain state regarding the consumption of alcoholic beverages consider three separate classes of persons with respect to drinking privileges: under the age of 19, ages 19 and 20, and age 21 or older. Thus it is always appropriate to ask, "What does 'equivalent' mean in this context?" whether we are considering mathematics or political philosophy or any other subject.

To obtain a precise definition of a broadly applicable sense of "equivalence," we consider an unspecified binary relation E on a set S and list those properties of relations that E should have to define a reasonable sense of "equivalence." To avoid using the word "equal" (which is reserved to mean "identical") in connection with E, we will read the symbol $a E b$ as "a is *equivalent* to b." The first property that we would insist upon is that each element of S should be equivalent to itself:

$a E a$ for all a in S.

This is the property of reflexivity listed in Table 2.1-1. The second condition that equivalence should satisfy is the following:

Whenever $a E b$, then $b E a$.

That is, E should be symmetric. The third property was recognized formally many centuries ago by Greek mathematicians: "Things equivalent to the same thing are equivalent to each other." If E is symmetric, this property can be assured by assuming further that E is transitive:

For all a, b, c in S whenever $a E b$ and $b E c$, then $a E c$.

Definition Let S be any nonempty set. An *equivalence relation E on S* is a binary relation that is reflexive, symmetric, and transitive.

EXAMPLES
1. Two plane triangles are congruent if and only if the measures of two sides and the included angle of one triangle are equal to the measures of two sides and the included angle of the other. Congruence is reflexive, symmetric, and transitive — essentially because of the way it is defined in terms of equality of measures. Thus congruence is an equivalence relation.

2. In the natural numbers "a divides b" is reflexive and transitive; it is not symmetric because 2 divides 4, but 4 does not divide 2. So it is *not* an equivalence relation on N.

3. The relation "is the same age as" is an equivalence relation.

4. If parallelism of lines in the plane is defined as "have no point in common," then parallelism is not reflexive. But if it is defined as "do not have a unique point of intersection," then parallelism is an equivalence relation. ∎

Each equivalence relation on S is defined in terms of one or more properties of the elements of S: age in Example 3, size and shape in Example 1. If we are interested only in people of different ages, it is convenient to group together all persons aged 0, all aged 1, and so on. In this way we could reduce the size of the set that we study — instead of all persons in China, for example, we could study a set of perhaps 120 age classes if longevity or some age-related characteristic was the point of our inquiry. We now turn our attention to that idea.

> **Definition** Let E be an equivalence relation on a nonempty set S. For each $a \in S$ we define the *E-equivalence class* $[a]$ by
>
> $$[a]_E = \{x \in S \mid x \, E \, a\};$$
>
> in words, $[a]_E$ denotes the set of all elements of S that are E-equivalent to a.

By definition each E-equivalence class $[a]_E$ of S is a subset of S. When only one equivalence relation E is being considered, no ambiguity is introduced by omitting the subscript E, thus denoting by $[a]$ the set of elements of S that are E-equivalent to a. Now we observe two properties of these equivalence classes.

1. For each $a \in S$ we have $a \in [a]$.

Property 1 is a restatement of the reflexive property of E; $a \, E \, a$, so $a \in [a]$.

2. Given any $a, b \in S$, either $[a] = [b]$ or $[a] \cap [b] = \varnothing$.

That is, any two equivalence classes are either *equal* or *disjoint*. In other words, property 2 states that if two equivalence classes have one common member, the two classes have all their members in common. Observe carefully that from $[a] = [b]$ we cannot conclude that $a = b$. For example, "is the same age as" is an equivalence relation on any nonempty set of people, and $[a] = [b]$ simply means that a and b have the same age; it does not mean that they are the same person.

To prove property 2, suppose that $[a] \cap [b] \neq \varnothing$. Then there exists an element x such that $x \in [a] \cap [b]$. That means that $x \ E \ a$ and $x \ E \ b$. Thus $b \ E \ x$ by symmetry of E, and from $b \ E \ x$ and $x \ E \ a$ and transitivity of E we have $b \ E \ a$, so $b \in [a]$. This fact enables us to show that $[b] \subseteq [a]$. To do so, let $y \in [b]$, which means that $y \ E \ b$. Again using transitivity of E and the fact that $b \in [a]$, we obtain $y \ E \ b$ and $b \ E \ a$, so $y \ E \ a$. Hence $y \in [a]$ for each $y \in [b]$, and we have $[b] \subseteq [a]$. Now if we return to the second sentence of this paragraph we see that the argument is symmetric in the symbols a and b. That is, we can repeat this argument with the roles of a and b interchanged, and thereby conclude that $[a] \subseteq [b]$. But now $[a]$ is a subset of $[b]$, and $[b]$ is a subset of $[a]$, so $[a] = [b]$. We have deduced that if $[a] \cap [b]$ is not empty then $[a] = [b]$, which is logically equivalent to property 2.

From properties 1 and 2 we deduce that each element of S belongs to one and only one E-equivalence class. Thus the effect on S of an equivalence relation E is to decompose S into a family $\mathscr{P}$ of nonempty and nonoverlapping subsets called E-equivalence classes having the property that the union of those classes is S.

We can represent the set S and the E-equivalence classes of S by a Venn diagram as shown in Figure 2.4-1. There the subsets labeled A, B, . . . , G represent the E-equivalence classes. Property 1 is exhibited by drawing the diagram so that the union of all equivalence classes covers all of S; that is, $A \cup B \cup \cdots \cup G = S$. Property 2 is indicated by drawing the equivalence classes so that no two classes overlap.

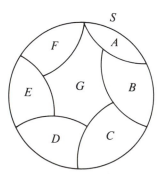

Figure 2.4-1

Definition   Let S be a set. A *partition* of S is a family $\mathscr{P}$ of nonempty subsets of S such that

1. The union of all the sets in the family $\mathcal{P}$ is S:

$$\bigcup_{A \in \mathcal{P}} A = S;$$

2. If $A, B \in \mathcal{P}$ and if $A \cap B \neq \varnothing$, then $A = B$.

The foregoing results can be summarized in this way:

If E is any equivalence relation on a set S, then the E-equivalence classes form a partition $\mathcal{P}_E$ of S.

EXAMPLE Consider S to be the set I of all integers and let m be any positive integer. For each pair a, b in I we say that

a is congruent to b modulo m if and only if $b - a$ is divisible by m. In symbols

$$a \equiv b \,(\text{mod } m) \Longleftrightarrow (b - a) \text{ is divisible by } m.$$

You may verify that congruence modulo m is reflexive, symmetric, and transitive and therefore is an equivalence relation on I. To see what the corresponding equivalence classes are, first recognize that when any integer c is divided by $m > 0$, there is a unique integer q (the quotient) and a unique integer r (the remainder) such that

$$c = mq + r \quad \text{and} \quad 0 \leq r < m.$$

(For example, if $c = 12$ and $m = 5$, then $12 = 5(2) + 2$; if $c = -12$ and $m = 5$, then $-12 = 5(-3) + 3$.) By definition $a \equiv b \,(\text{mod } m)$ is valid if and only if m divides $b - a$ with 0 remainder, and the latter occurs if and only if a has the same remainder as b when each is divided by m. In this sense there are exactly m distinct types of integers, in accordance with the m possible remainders $0, 1, 2, \ldots, m - 1$ when an integer is divided by m. Thus the equivalence classes are of the form

$$[k]_m = \{x \in I \mid x \text{ has remainder } k \text{ when divided by } m\}$$

for $k = 0, 1, \ldots, m - 1$.

To illustrate this description of the equivalence classes, consider the case where $m = 5$. Then

$$[0]_5 = \{\ldots, -10, -5, 0, 5, 10, \ldots\},$$
$$[1]_5 = \{\ldots, -9, -4, 1, 6, 11, \ldots\},$$
$$[2]_5 = \{\ldots, -8, -3, 2, 7, 12, \ldots\},$$

$$[3]_5 = \{. \ . \ . \ , -7, -2, 3, 8, 13, . \ . \ .\},$$
$$[4]_5 = \{. \ . \ . \ , -6, -1, 4, 9, 14, . \ . \ .\}.$$

Observe that each integer is in one and only one of these $m = 5$ equivalence classes. Observe also that any two integers in a single equivalence class have the same remainder when each is divided by 5. For example, the remainders of 8 and -7, when divided by 5, are both 3 because $8 = 5(1) + 3$ and $-7 = 5(-2) + 3$. (Remember that in this example the remainder r must satisfy $0 \leq r < 5$.) ∎

Having seen that each equivalence relation on S determines a partition of S, we now show that each partition of S determines an equivalence relation.

Let $\mathscr{P}$ be any partition of S. For each $a \in S$ let $[a]_\mathscr{P}$ denote the unique subset of that partition to which a belongs. We denote by $E_\mathscr{P}$ the binary relation defined on S by the rule

$$a \ E_\mathscr{P} \ b \quad \text{if and only if} \quad a \in [b]_\mathscr{P}.$$

Then $E_\mathscr{P}$ is an equivalence relation on S.

To justify this assertion we need to show that the relation $E_\mathscr{P}$ is reflexive, symmetric, and transitive.

1. Reflexive: $a \ E_\mathscr{P} \ a$ because $a \in [a]_\mathscr{P}$.
2. Symmetric: If $a \ E_\mathscr{P} \ b$, then $a \in [b]_\mathscr{P}$. Then $[b]_\mathscr{P}$ is a member of the family $\mathscr{P}$ of subsets of S, and both a and b are in $[b]_\mathscr{P}$. But only one such subset contains a and only one contains b, so $[a]_\mathscr{P} = [b]_\mathscr{P}$. This means that $b \in [a]_\mathscr{P}$ and hence $b \ E_\mathscr{P} \ a$.
3. Transitive: Let $a \ E_\mathscr{P} \ b$ and $b \ E_\mathscr{P} \ c$. Then $c \ E_\mathscr{P} \ b$ by symmetry, so a and c both belong to $[b]_\mathscr{P}$, which assures that $[a]_\mathscr{P} = [c]_\mathscr{P}$ and therefore $a \ E_\mathscr{P} \ c$.

At this point you will find it useful to reread the first page of this section, observing that our initial examples of equivalence relations were described in terms of partitions. To summarize, every method of assigning each member of a set to exactly one of a list of classifications defines a partition of the set and thus defines an equivalence relation on the set. Conversely, any binary relation on the set that is reflexive, symmetric, and transitive is an equivalence relation, and the corresponding equivalence classes decompose the set into disjoint subsets whose union is the entire set.

Exercises 2.4

1. Let S be the set of all students enrolled at College C in a given semester. Which of the following binary relations on S are equivalence relations, and which are not? Explain.

   (a) $x R y$ means that x lives in the same building as y.
   (b) $x R y$ means that x weighs more than y.
   (c) $x R y$ means that x is enrolled in exactly two classes in which y is enrolled. (Assume each student is enrolled in four classes.)
   (d) $x R y$ means that the faculty advisor of x is also the faculty advisor of y. (Assume each student has one and only one faculty advisor.)
   (e) $x R y$ means that x's numerical grade on the mathematics placement test minus y's grade on that test is divisible by 3. (Assume that each student has taken the mathematics placement test.)

2. Let p be a fixed point in a plane and L a fixed line. Let S be the set of all points in the plane. Which of the following binary relations on S are equivalence relations, and which are not? Explain.

   (a) $x R y$ means that x and y are equidistant from p.
   (b) $x R y$ means that there exists a line passing through the points x, y, p.
   (c) $x R y$ means that x and y lie on a line perpendicular to L.
   (d) $x R y$ means that x and y are equidistant from the line L.
   (e) $x R y$ means that the distance between x and y is less than 2 centimeters.

3. For each of the relations in Exercise 2 that is an equivalence relation, give a geometric description of the equivalence classes. In each case draw a picture showing three distinct equivalence classes.

4. Let R be a binary relation on a set S. Let G be the graph representing R, and let M be the 0-1 matrix representing R. What special property does G have and what special property does M have when

   (a) R is reflexive?
   (b) R is symmetric?
   (c) R is transitive?

5. Let R be a binary relation on a set. Complete each of the following.

   (a) R is not reflexive if and only if _____.
   (b) R is not symmetric if and only if _____.
   (c) R is not transitive if and only if _____.

6. Give four examples of equivalence relations not mentioned in this section or this set of exercises. Explain why your examples are equivalence relations.

7. **(a)** Let m be any positive integer. Prove that congruence modulo m is an equivalence relation on the set I of all integers.
   (b) List six distinct elements, three negative and three positive, in each of the five distinct equivalence classes for the equivalence relation congruence mod 5.
   (c) Determine three distinct integers, x, y, w such that $[x]_5 = [y]_5 = [w]_5$.

8. Which of the following relations on the set I of integers are equivalence relations, and which are not? Explain.
   (a) $x R y$ if and only if $x + y \equiv 0 \pmod 2$.
   (b) $x R y$ if and only if $x + y \equiv 0 \pmod 3$.
   (c) $x R y$ if and only if $x^2 \equiv y^2 \pmod 3$.

9. For each of the relations in Exercise 8 that is an equivalence relation, list four distinct members of the equivalence class containing the integer 1.

10. As usual, let Q be the set of rational numbers and I the set of integers.
    (a) If E is the binary relation defined on Q by

    $$x E y \quad \text{if and only if} \quad x - y \in I,$$

    verify that E is an equivalence relation on Q.
    (b) Let E be the equivalence relation defined in (a). List four members in each of the equivalence classes $[\frac{2}{3}]$ and $[401]$ and then describe how one can generate any number of distinct members in each of these classes.
    (c) If $a \in Q$, we define $\text{INT}(a)$ to be the largest integer less than or equal to a. If R is the binary relation defined on Q by

    $$x R y \quad \text{if and only if} \quad \text{INT}(x) = \text{INT}(y),$$

    verify that R is an equivalence relation on Q.
    (d) Do part **b** with E replaced by the equivalence relation R as defined in **c**.

11. Let S denote a set of seven students who are studying in various universities throughout the world. Let R be a binary relation defined on S by

    $$x R y \quad \text{if and only if} \quad x \neq y \text{ and } x \text{ can send a message directly to } y.$$

The 0-1 matrix for R is given by

$$M = \begin{array}{c} \\ A \\ B \\ C \\ D \\ E \\ F \\ G \end{array} \begin{array}{ccccccc} A & B & C & D & E & F & G \\ \left(0 \right. & 1 & 0 & 0 & 1 & 0 & 0 \\ 0 & 0 & 1 & 1 & 0 & 0 & 1 \\ 0 & 0 & 0 & 0 & 0 & 0 & 0 \\ 0 & 1 & 1 & 0 & 0 & 0 & 1 \\ 1 & 0 & 0 & 1 & 0 & 0 & 0 \\ 0 & 0 & 0 & 0 & 1 & 0 & 0 \\ 0 & 1 & 0 & 1 & 0 & 0 & \left. 0 \right) \end{array}.$$

(a) A second binary relation T can be defined on S by

> $x \, T \, y$ if and only if $x = y$, or $x \neq y$ and x can send a message to y directly or indirectly through one or more intermediate persons.

Show that T is reflexive and transitive.

(b) A third binary relation E can be defined on S by

> $x \, E \, y$ if and only if $(x \, T \, y)$ and $(y \, T \, x)$.

Show that E is an equivalence relation on S.

(c) Determine the E-equivalence classes of S and interpret the meaning of those classes.

12. Let R be a symmetric and transitive relation on a set S. Furthermore, suppose that for every $x \in S$ there is an element $y \in S$ such that $x \, R \, y$. Show that R is an equivalence relation on S.

13. For each number a in the set Q of all rational numbers, let $S(a)$ be defined by

$$S(a) = \{ x \in Q \mid x = a + n \text{ for some } n \in I \}.$$

Let $\mathscr{P}$ denote the family of subsets of Q defined by

$$\mathscr{P} = \{ S(a) \mid a \in Q \text{ and } 0 \leq a < 1 \}.$$

(a) Verify that $\mathscr{P}$ is a partition of Q.

(b) Is $\mathscr{P}$ the family of all equivalence classes for either of the two relations E or R as defined in Exercise 10? Justify your answer.

2.5 Partial Order Relations

PROBLEM The Goormay family was trying to decide how to choose among six standard brands of peanut butter. The teenager wanted the brand that tasted best. One parent proposed that nutrition should be the principal criterion, and the other parent was most concerned with cost. They then agreed upon the rankings shown in Table 2.5-1, where 1 denotes best taste, greatest nutrition, and lowest price per ounce. Which brand should the family choose?

Brand	Taste	Nutrition	Price
A	2	1	6
B	3	2	1
C	1	3	3
D	6	4	4
E	4	5	2
F	5	6	5

Table 2.5-1   ∎

To determine a composite choice, the family agreed that brand X should be regarded as preferable to brand Y if and only if brand X was ranked above brand Y in each of the three criteria for comparison shown in Table 2.5-1. Thus

A, D, and F are not preferred to any brand (because each is ranked lowest in some category).

No brand is ranked higher than A, B, and C (because each is ranked highest in some category).

B is preferred to D, E, and F.

C is preferred to D and F.

E is preferred to F.

This composite ranking information can be presented more clearly in a diagram as shown in Figure 2.5-1, where each brand is represented by a labeled point, and a directed segment is drawn from point X to point Y if and only if

Y is preferred to X, *and*

no point Z exists such that Y is preferred to Z and Z is preferred to X.

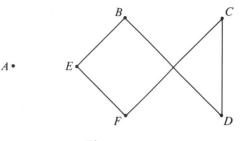

Figure 2.5-1

Such a representation is called a *Hasse diagram.* By placing the labeled points suitably on the plane, we can adopt the convention that each directed segment is directed *upward.*

Although this problem is posed in apparently trivial terms, it is a simple example of the very significant problem of social choice: Given a preferential ranking of three or more options by each individual in a group of two or more persons, which single choice for the group best represents the individual preferences of the members?

In this example the rule for obtaining a group comparison of brand X and brand Y from three individual comparisons seems reasonable, but it did not produce a strong total order relation among the six brands of peanut butter, even though each member of the Goormay family had submitted a strong total ordering of the six brands; that is, each individual comparison was irreflexive, asymmetric, and transitive, and also possessed the trichotomy property, as defined in Section 1.9 for the set N of natural numbers. The group comparison turned out to be irreflexive, asymmetric, and transitive, but it failed to satisfy trichotomy in that it did not provide a comparison between all pairs of brands, although it did compare some pairs of brands. Thus the group comparison rule defined a strong partial ordering (but not a strong total ordering) of the six brands being considered.

Another example of a strong partial ordering can be defined on the family $\mathcal{F}$ of all subsets of a given set S by the binary relation $\subset$, is a proper subset of. On the other hand, the binary relation $\subseteq$, is a subset of, does not define a strong partial ordering on $\mathcal{F}$ because it fails to be irreflexive and asymmetric; instead, it is reflexive, antisymmetric, and transitive. Such an ordering is called a *weak* partial ordering. We now state formal definitions of these concepts for any nonempty set.

> **Definition** Let S denote a nonempty set and let P denote a binary relation on S. If P is irreflexive, asymmetric, and transitive, P is called a *strong partial ordering of S.* If P is reflexive, antisymmetric, and transitive, P is called a *weak partial ordering of S.*

Notice that the distinction between these two types of orderings is analogous to both the familiar distinction between the two types of numerical inequality,

$$x < y \quad \text{and} \quad x \leq y,$$

and the distinction between the two types of subset inclusion,

$$A \subset B \quad \text{and} \quad A \subseteq B.$$

Having made the distinction between a strong partial ordering and a weak partial ordering of S, we point out that in Exercise 9 you are asked to show that if P_1 is a strong partial ordering of S, then a weak partial ordering P_2 can be defined on S by the rule

$$x \, P_2 \, y \quad \text{if and only if} \quad \text{either } x = y \text{ or } x \, P_1 \, y.$$

Conversely, if P_2 is any weak partial ordering of S, then a strong partial ordering P_1 can be defined on S by the rule

$$x \, P_1 \, y \quad \text{if and only if} \quad \text{both } x \, P_2 \, y \text{ and } x \neq y.$$

Consequently, the distinction between strong and weak partial order relations frequently can be ignored, thereby simplifying our language without introducing ambiguity.

> **Definition** A *partially ordered set* consists of a nonempty set S on which is defined a strong or weak partial order relation P. S is said to be *partially ordered by P*. We also say that S is partially ordered *relative to P*.

EXAMPLES OF PARTIALLY ORDERED SETS

1. Let $S = \{a, b, c, d, e\}$ and let $\mathcal{F}$ denote the family of all subsets T of S such that either $\{a, b, c, d\}$ or $\{c, d, e\}$ is a subset of T. $\mathcal{F}$ is partially ordered by the subset relation and has five members: S itself and

$$1\{a, b, c, d\}; \quad 2\{c, d, e\}; \quad 3\{c, d, e, a\}; \quad 4\{c, d, e, b\}.$$

A Hasse diagram of $\mathcal{F}$ is given in Figure 2.5-2.

2. The organizational chart of a corporate administration defines the principal lines of authority within the corporation, and it can be interpreted as a Hasse diagram of the administrative positions relative to the relation "responsibility for." The occupant of each position on the

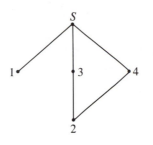

Figure 2.5-2

chart is responsible for that position and for all positions that can be reached by moving downward from the given position along lines of the chart.

3. On each day of the baseball season newspapers report the standings of the 26 teams in the major leagues. The National League has 12 teams separated into 2 divisions of 6 teams each, whereas the American League consists of 14 teams, 7 in each of 2 divisions. The daily rankings are given by means of separate listings of all teams in each of the 4 divisions. The ranking within each division is a total ordering of the teams in that division. Taken together, the 4 lists of rankings then comprise a partial ordering of the 26 teams in the major leagues, but that partial ordering fails to provide an official comparison of teams in different divisions.

4. The concept of divisibility defines a partial ordering on the set of positive natural numbers. We write $a\ D\ b$ and say that "a divides b" if and only if $b = ac$ for some positive natural number c. Figure 2.5-3 shows a fragment (for $a \le 10$ and $b \le 20$) of the Hasse diagram of the positive natural numbers relative to the relation D.

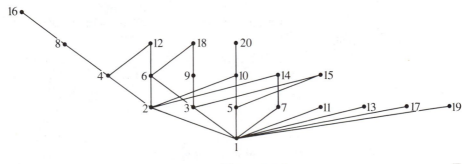

Figure 2.5-3

Let us next consider the concept of total ordering that was described for N in Section 1.9 to extend that type of relation from the set of natural numbers to nonempty sets in general.

Definition   Let S be a nonempty set partially ordered relative to a relation P. S is said to be *totally ordered* if and only if P satisfies this additional condition:

For all a, $b \in S$ either $a\,P\,b$ or $b\,P\,a$ or $a = b$.

When S is totally ordered by P, the relation P is called either a *weak total ordering* or a *strong total ordering,* according as P is a weak or a strong partial ordering. A totally ordered set is also said to be *linearly ordered.*

Notice that in a set S that is totally ordered by P, any two distinct elements are related: either $a\,P\,b$ or $b\,P\,a$. In a partially ordered set this need not be the case. In the event that P is a strong total ordering, the additional condition in the preceding definition is equivalent to the assertion that for all a and b in S one and only one of the following statements is true:

$$a\,P\,b; \quad b\,P\,a; \quad a = b.$$

This assertion is the *trichotomy property,* as defined in Section 1.9 for the set N of natural numbers. If the relation P is a weak total ordering, trichotomy does not hold because all three of the statements are true when $a = b$.

EXAMPLES OF TOTALLY ORDERED SETS

5. The set R of real numbers is totally ordered by the "less than or equal to" relation $\leq$. You should verify this. You should also convince yourself that the "less than" relation $<$ is a strong partial order relation that relates each pair of distinct real numbers in precisely one order, and thus it is a strong total order relation.

6. The order of letters in the alphabet is used to totally order the set of all words, where a "word" is any finite sequence of letters, such as mnbvcxzasdfgtgbfe. Any two different words can be placed in lexicographic order by comparing them letter by letter from left to right until two letters in the same position are different. The alphabet, extended by adjoining a blank symbol, which is placed in front of the standard alphabet, determines which of the two words precedes the other. For example, consider how the words "there," "they," and "the" are al-

phabetized. All three words coincide for the first three letters. The fourth symbols are *r*, *y*, and "blank," respectively. Hence "the" comes first, "there" is next, and "they" comes last. ∎

A Fundamental Property of Partially Ordered Sets

Partially ordered sets are very general mathematical systems, because each consists only of a nonempty set *S* and a partial order relation *P* defined on *S*. Any two elements *a* and *b* of *S* may or may not be related by *P*, so our main interest in the structure of a given partially ordered set focuses on two types of subsets of *S* that are quite opposite in character relative to *P*:

1. Subsets called *chains* that are totally ordered by *P*
2. Subsets called *antichains* that are completely unordered by *P*, that is, no two distinct elements of an antichain are related by *P*.

From the definition of a totally ordered set, we see that each one-element subset of *S* is a chain. But each one-element subset of *S* also is an antichain, because the defining condition of an antichain is vacuously satisfied. That is, it is true that no two distinct elements of a one-element set are related by *P*, because there are not two distinct elements in such a subset. A subset having more than one member cannot be both a chain and an antichain, and it might be neither. The following examples will help to clarify the concepts of chain and antichain.

EXAMPLES 7. Consider the seven-element partially ordered set represented by the Hasse diagram in Figure 2.5-4. There are many chains because any one-element set is a chain and any nonempty subset of a chain is a chain. The longest chain has four elements, and there are three such chains: *ACDE*, *FGDE*, and *FCDE*. Any three-element chain is a subset

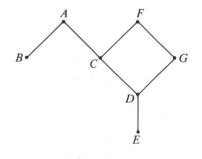

Figure 2.5-4

of some four-element chain. Any two-element chain is either *AB* or a subset of some three-element chain. There are many antichains also: Any one-member set is an antichain, and any pair of distinct unrelated elements is an antichain; for example, *AF, AG, BC, BD, BE, BF, BG,* and *CG. BCG* is the largest antichain, and no other antichain is as large. However, the set *ABC* is neither a chain nor an antichain.

8. In Example 3 each of the 4 divisions of the major leagues of baseball forms a chain, and those 4 chains partition the set of all teams in the major leagues. (The longest chain contains 7 teams, from which we can easily deduce that there are at most $4(7) = 28$ teams in the two leagues.) The set of all teams can be partitioned into chains in many other ways, but each such partition must contain at least 4 chains, because each of the 4 divisions is a chain, and no team belongs to more than 1 division. ∎

Any antichain is a nonempty set of teams, no two of which are in the same division. Thus four is also the largest number of teams in any antichain. See Figure 2.5-5.

In Example 7, we observe that the partially ordered set in Figure 2.5-4 can be partitioned into as few as three chains: *FGDE, AC,* and *B,* for example. Of course, three is the smallest possible number of chains in a chain partition of that set, because the number of chains in any chain partition must be at least as large as the number of elements in the largest antichain. We shall now see that these observations apply without change in any partially ordered set.

Definition A *chain decomposition C* of a partially ordered set *S* is a partition of *S* into subsets, each of which is a chain.

Let *S* denote any partially ordered set, *C* any chain decomposition of *S*, and *A* any antichain of *S*. We want to investigate the relation between the

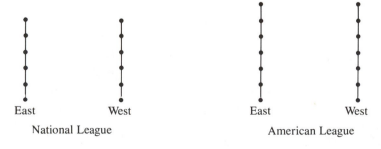

East   West   East   West

National League   American League

Figure 2.5-5

number $n(A)$ of elements in the antichain A and the number $n(C)$ of chains in the chain decomposition C. Because

1. no two distinct elements in A are related by the partial order relation, and
2. each element of S belongs to exactly one chain of C,

the inequality

$$n(A) \leq n(C)$$

holds for *every* antichain A and *every* chain decomposition C of S. If we hold C fixed temporarily and consider all antichains A, we obtain the following stronger statement:

$$\underset{\text{all } A}{\text{maximum}} \, [n(A)] \leq n(C) \qquad \text{for every chain decomposition } C.$$

Now let C vary over all chain decompositions of S; because the previous inequality is valid for every C, the smallest value of $n(C)$ must also satisfy that inequality. Let a and c be the positive integers defined by

$$a = \underset{\text{all } A}{\max} \, n(A) \quad \text{and} \quad c = \underset{\text{all } C}{\min} \, n(C);$$

then we conclude that $a \leq c$.

By a more complicated argument the opposite inequality can also be proved, thereby providing a fundamental and widely applicable theorem about the structure of partially ordered sets, as published by Robert P. Dilworth in 1950.

Theorem 2.5-1 *(Dilworth's theorem)* Let S be any partially ordered set. The smallest number of chains in any chain decomposition of S equals the largest number of elements in any antichain of S.

APPLICATION Each July 4 the Miller family gathers at the old homestead for a reunion. Millers of all ages come from miles around for a feast, games, and news of relatives. At the most recent Miller reunion one of the younger Millers, who happened to know some mathematics, counted 101 genetic descendants of Isaac Miller at the reunion. Then she promptly announced,

Among those present today are 2 genetic descendants of Isaac Miller, 1 of whom is a great-great-great-grandparent of the other; or else there are 21 Miller descendants present, none of whom is descended from any of the other 20.

She was correct, and at the end of the day all the Millers returned home full of happy memories of the day and new-found respect for young cousin Elsie and her impressive knowledge of the Miller family tree. Did Elsie need to know anything about the family tree to make her assertion? (See Exercise 7.)   ■

Exercises 2.5

1. Let $\mathscr{F}$ be the family of all subsets of a nonempty set S.
   (a) Verify that $\mathscr{F}$ is partially ordered by the subset relation $\subseteq$.
   (b) If $\mathscr{F}$ is totally ordered by $\subseteq$, what can you conclude about the size of S?

2. Let $\mathscr{F}$ be the family of all subsets of a set S, and let R denote the binary relation defined on $\mathscr{F}$ by the rule

   $A \, R \, B$ if and only if $A = B$ or A is not a subset of B.

   Is R a partial ordering of $\mathscr{F}$? Explain.

3. Let $S = \{2, 3, 5, 6, 10, 20, 24, 30, 36, 40, 48\}$. Let D be the binary relation defined on S by

   $a \, D \, b$ if and only if a divides b.

   (a) Prove that D is a partial ordering of S.
   (b) Is D a total ordering of S? Explain.
   (c) Draw a Hasse diagram for the set S partially ordered by D.
   (d) Determine the 0-1 matrix that represents D.

4. Let $\mathscr{F}$ denote the family of all subsets of the set $\{1, 2, 3\}$, partially ordered by the proper subset relation $\subset$.
   (a) List each member of $\mathscr{F}$ and label each such subset with a letter from A through H.
   (b) Draw a Hasse diagram of the partially ordered set $\mathscr{F}$.
   (c) List all chains of $\mathscr{F}$ having three or more members.
   (d) Write a minimal chain decomposition of $\mathscr{F}$. How can you be certain that your chain decomposition is minimal?

5. Let S and D be defined as in Exercise 3.
   (a) List all chains of three or more elements in the partially ordered set S.

   (b) List all antichains of S containing three or more elements.

   (c) Determine a chain decomposition of S consisting of five chains.

   (d) Use Dilworth's theorem to determine the minimal number of chains needed to obtain a chain decomposition of S.

   (e) Determine two chain decompositions of S containing a minimal number of chains.

6. Illustrate Theorem 2.5-1 for the partially ordered set represented by Figure 2.5-3. Select a maximal antichain A and a minimal chain decomposition C, and state clearly how you are sure that A is maximal and C is minimal.

7. In the application that concludes this section, let S denote the set of all genetic descendants of Isaac Miller present at the Miller reunion. Define a binary relation R on S by the rule that $x \, R \, y$ if and only if x is a genetic descendant of y.

   (a) Show that R is a strong partial ordering of S.

   (b) An antichain of S is either a one-person subset of S or a subset of S containing more than one person, none a descendant of any other person in that subset. If the size of every antichain is less than 21, what can you conclude from Dilworth's theorem about a chain decomposition of S? Explain fully.

   (c) Answer the question asked in the final sentence of the application, explaining your answer in detail.

8. Let A be the set of all natural numbers that can be expressed as an integral power of 2, and let R be the binary relation on A defined by the rule

$$a \, R \, b \quad \text{if and only if} \quad a \neq b \text{ and } a \text{ divides } b.$$

Is R a strong total order relation on A? Show your reasoning.

9. (a) Let S denote a nonempty set, and let R denote a strong partial order relation on S. Let P denote the relation obtained from R by the rule that for all a and b in S

$$a \, P \, b \quad \text{if and only if} \quad \text{either } a \, R \, b \text{ or } a = b.$$

Prove that P is reflexive, antisymmetric, and transitive, and hence is a weak partial order relation on S.

   (b) Now let P denote any weak partial order relation on S, and let R denote the binary relation on S obtained from P by the rule that for all a and b in S

$$a \, R \, b \quad \text{if and only if} \quad a \, P \, b \text{ and } a \neq b.$$

Prove that R is irreflexive, asymmetric, and transitive, and hence is a strong partial order relation on S.

10. Let R be a binary relation on the cartesian product $N \times N$, defined by

$$(a, b) \: R \: (c, d) \quad \text{if and only if} \quad a < c \text{ and } b < d.$$

Is R a strong total order relation on $N \times N$? If your answer is yes, prove your assertion. If your answer is no, which of the four properties of a strong total order relation are satisfied, and which are not satisfied?

11. In the problem posed at the start of this section, which brand of peanut butter should the family choose and why? Defend your answer, and also explain why some family members might not agree with your choice.

2.6 Functions and Binary Operations

You have already worked with some numerical-valued functions, such as

$$f(x) = x^2 - 4x + 3.$$

In this case the function f associates with each real number x the real number $f(x)$ obtained by squaring x, subtracting $4x$, and adding 3. The particular formula for calculating the number $f(x)$ in terms of the number x is not important here. We want to emphasize only that this function associates with each real number x one specific real number $f(x)$. In this example the domain of f is the set of all real numbers, and the range of f is the set of all real numbers $y \geq -1$.

Because mathematics uses both numerical and nonnumerical functions let us now consider functions generally, from any nonempty set A to any nonempty set B.

> **Definition** Let A and B be nonempty sets. A *function* (or *mapping*) f from A into B is a binary relation from A to B having the special property that
>
> $$\text{if } a \, f \, b \text{ and } a \, f \, c, \text{ then } b = c.$$

This condition simply specifies that no element of A can be related by f to more than one element in B. Thus a relation that is a function associates with each element in its domain *one and only one* element in its range. (The domain and range of a binary relation were defined in Section 2.1.) We can describe this idea pictorially by recalling that a binary relation can be sketched as in Figure 2.1-1. Such a relation is a function if and only if *at most one* arrow is drawn from any element of *A*. Equivalently, from each element *a* in the domain of *f* exactly one arrow is drawn, and it ends at the one element *b* in the range of *f* such that *a f b*.

Because *b* is determined uniquely by *f* and *a*, we use the usual notation for functions and write

$$b = f(a) \quad \text{instead of} \quad a f b.$$

The element $f(a)$ of *B* is called "the value of *f* at *a*," and the association that defines *f* can be written

$$f: a \rightarrow f(a)$$

and can be read "*f* maps *a* to $f(a)$."

If each element of *B* is in the range of *f*, then *f* is said to be a function from *A onto B*.

It is sometimes useful to think of a function as a machine, specifically designed to receive any element *a* of the domain of *f* as input and to deliver the associated member $f(a)$ of the set *B* as output. See Figure 2.6-1.

EXAMPLES OF FUNCTIONS

1. The relation "is the mother of" is not a function because many women have more than one child. However, "is the offspring of" is a function from the set *A* of all persons into the set *B* of all female persons of the past or present, because each person has one and only one biological mother. The domain is the set of all persons, past or present, and the range is the set of all females who have had one or more children. If the

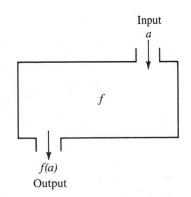

Input
a

f

f(a)
Output

Figure 2.6-1 *A function machine.*

range were not restricted to be a subset of females, then "is the offspring of" is not a function because each of us has two biological parents.

2. The sentence, "For each natural number x, $p(x)$ is the smallest prime number larger than x," defines a function p from N into N because for each natural number x there is one and only one "smallest prime larger than x." The fact that we have no formula for calculating $p(x)$ from x is not relevant to the question of whether that sentence defines a function. The function p is not onto N because not every natural number is prime.

3. With each subset A of a nonempty set S there is an associated function whose domain is S and whose range is a subset of the two-element set $\{0, 1\}$. This function is denoted I_A and is called the *characteristic* function of A or the *indicator* function of A. It is defined for each $x \in S$ by the compound rule

$$I_A(x) = \begin{cases} 1 & \text{if } x \in A, \\ 0 & \text{if } x \notin A. \end{cases}$$

Such functions represent the various subsets of S numerically so that the algebra of subsets can be performed by calculating with the associated characteristic functions. For example, let A and B be subsets of S, and as usual let A' denote the complement of A in S. Then the following properties hold for all x in S.

(a) $I_{A'}(x) = 1 - I_A(x)$.
(b) $I_{A \cap B}(x) = I_A(x)I_B(x)$.
(c) $I_{A \cup B}(x) = I_A(x) + I_B(x) - I_{A \cap B}(x)$.
(d) If $A = \{x_1, x_2, \ldots, x_m\}$, where $x_i \neq x_j$ whenever $i \neq j$, then

$$n(A) = \sum_{k=1}^{m} I_A(x_k) = I_A(x_1) + I_A(x_2) + \cdots + I_A(x_m)$$

$$= \sum_{x \in S} I_A(x).$$

4. A function s having domain N and range a subset of R is called a real-valued *sequence*. Instead of writing the terms of the sequence in function notation as $s(k)$, we usually use k as a subscript, writing the terms s_0, s_1, s_2, and so on. In that notation the symbol $\{s_k\}$ refers to the entire sequence, and s_k denotes the kth term of the sequence. A sequence is said to be finite if and only if $s_k \neq 0$ for only a finite number of values of k. Otherwise, the sequence is infinite.

5. The process of assigning a Social Security number to each applicant is a continuing procedure for defining a function from a subset A of all people to the set B of all nine-digit numerals. This function is supposed

to have a very special property—a different number is assigned to each applicant. That is, no two persons should be assigned the same Social Security number. ∎

Any function assigns a unique value $f(x)$ to each x in its domain. But in Example 5 we would like to be sure that for all x and y in the domain of f

$$\text{if } x \neq y \qquad \text{then } f(x) \neq f(y).$$

Any such function is said to be *one-to-one,* or *reversible,* or is said to define a *one-to-one correspondence* between the elements of its domain and the elements of its range. In Section 1.4 we saw that the concept of a one-to-one correspondence is fundamental to the process of counting, and indeed it is basic throughout mathematics. To see why the term "reversible" is appropriate, consider the pictorial representation of a binary relation using arrows from A to B as described in Section 2.1.

The three diagrams in Figure 2.6-2 emphasize the following distinctions.

1. The diagram of a *relation* may have *many* arrows leaving each element of its domain and *many* arrows terminating at each element of its range.

2. The diagram of a *function* must have *precisely one* arrow leaving each element of its domain and may have *many* arrows terminating at each element of its range.

3. The diagram of a *one-to-one function* must have *precisely one* arrow leaving each element of its domain and *precisely one* arrow terminating at each element of its range.

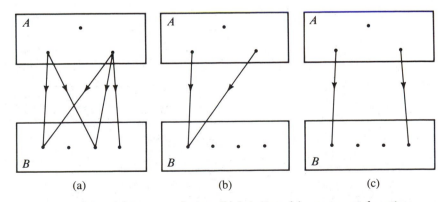

(a) (b) (c)

Figure 2.6-2 *(a) Binary relation; (b) function; (c) one-to-one function.*

Only in the case of a one-to-one function is it possible to reverse the direction of each arrow (without otherwise changing the diagram) and thereby obtain a function g from B to A. In that case g is also one-to-one, and g reverses the function f in the sense that

$$g(b) = a \quad \text{if and only if} \quad b = f(a)$$

for every a in the domain of f and every b in the range of f. It follows that

$$g(f(a)) = a \quad \text{and} \quad f(g(b)) = b$$

for every a in the domain of f and every b in the range of f. Note that the domain of g is the range of f, and the range of g is the domain of f. The function g is called the *inverse* of f, and f is called the *inverse* of g.

The displayed equations in the previous paragraph illustrate a special case of the *composition* of two functions, or the successive application of two mappings, an important way of combining two functions, f and then g, to obtain a third function, denoted $g \circ f$. To describe this concept informally, we use the function machine analogy as in Figure 2.6-1. Imagine that we have two function machines; one (called f) will accept as input any element x from its domain and will deliver as output an element $f(x)$ of its range. Now suppose that our other function machine (called g) will accept as input any element delivered as output from the f machine. That is, suppose that the range of f is a subset of the domain of g. Then we can take the output $f(x)$ of the first machine and introduce it as input to the second machine to obtain the element $g(f(x))$ as output of the second machine, as in Figure 2.6-3.

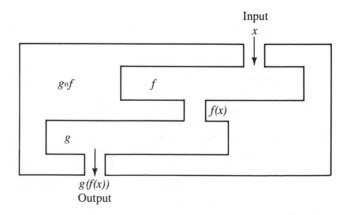

Figure 2.6-3

Definition Let f be a function from A into B, and let g be a function from B into C such that the domain of g includes the range of f. Then the *composition* $g \circ f$ of f by g is the function from A into C defined by

$$(g \circ f)(x) = g(f(x)) \qquad \text{for each } x \text{ in the domain of } f.$$

Observe that $f \circ g$ is not necessarily defined even though $g \circ f$ is defined, because for $f \circ g$ to be defined the domain of f must contain the range of g. The domain of f is a subset of A, whereas the range of g is a subset of C.

We can also use a mapping diagram as in Figure 2.6-4 rather than the function machine analogy to convey the concept of the composition of functions.

As an example of numerical functions, let

$$f(x) = x + 3, \qquad g(x) = x^2.$$

In this case both $g \circ f$ and $f \circ g$ are defined, but not equal, because

$$(g \circ f)(x) = g(f(x)) = g(x + 3)$$
$$= (x + 3)^2 = x^2 + 6x + 9,$$

whereas

$$(f \circ g)(x) = f(g(x)) = f(x^2) = x^2 + 3.$$

With the concept and notation of functions clearly in mind, we can now define what is meant by a *binary operation* on a set S. In so doing we want to capture the essence of such familiar operations as addition of numbers $(x + y)$, intersection of sets $(A \cap B)$, disjunction of statements $(p \lor q)$, and composition of functions $(g \circ f)$. Each of these operations combines two elements of a given set to obtain another element of that set.

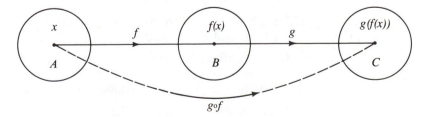

Figure 2.6-4

> ***Definition*** Let S be a nonempty set. A *binary operation* $*$ on S is a function with domain $S \times S$ and range a subset of S.

First, let us consider notation: Instead of using function notation for binary operations, we shall adopt a notation like that used for such familiar binary operations as arithmetic addition and multiplication and set union and intersection. Thus we shall write $a * b$ to denote the result of applying the binary operation $*$ to the ordered pair (a, b) in $S \times S$.

Observe carefully that $a * b$ must be defined for all a and all b in S, and, furthermore, $a * b$ must be in S for all a, b in S. Therefore, subtraction is not a binary operation on N because $3 - 7$ (for example) is not an element of N. If $*$ is a binary operation on S as defined above, S is said to be *closed* relative to $*$, which means simply that whenever a, b are in S, then $a * b$ also is in S.

EXAMPLES OF BINARY OPERATIONS

6. Subtraction is a binary operation on the set I of all integers. But division is not an operation on I. Strictly speaking, division is not an operation on the rational numbers or the real numbers because division by 0 is not defined. However, division is an operation on the set of all nonzero rational numbers and similarly on the set of all nonzero real numbers.

7. Set union and intersection are binary operations on the family $\mathcal{F}$ of all subsets of a nonvoid set S. The complement operation in $\mathcal{F}$ is a function whose domain is $\mathcal{F}$ and whose range is $\mathcal{F}$. It is called a *unary operation* on $\mathcal{F}$ because it maps each element of $\mathcal{F}$ into some member of $\mathcal{F}$. ∎

Exercises 2.6

1. Let f be the function from R into R defined by $f(x) = 3x - 2$.
   (a) Find $f(0)$, $f(-\frac{1}{2})$, $f(\frac{1}{2})$.
   (b) Show that f is one-to-one.
   (c) Show that f is a function from R onto R.

2. Let g be a function from R into P, where P is the set of positive real numbers and g is defined by $g(x) = x^2 + 1$.
   (a) Find $g(0)$, $g(-\frac{1}{2})$, $g(\sqrt{2})$.
   (b) Show that g is not one-to-one.
   (c) Show that g is not a function from R onto P.

3. Let f and g be defined as in Exercises 1 and 2.
   (a) Find $(g \circ f)(-1)$, $(g \circ f)(0)$, $(g \circ f)(1)$.

(b) Find $(f \circ g)(-1)$, $(f \circ g)(0)$, $(f \circ g)(1)$.

(c) Find equations defining the two functions, $g \circ f$ and $f \circ g$.

4. Let A be the set of all persons, past or present. Which of the following statements define functions f from A into A? For those that are functions, describe the domain and range. Explain your answers.

(a) $f(x)$ is the grandmother of x.

(b) $f(x)$ is the paternal grandmother of x.

(c) $f(x)$ is the youngest sister of x.

5. Let f and g be functions from R into R, defined by

$$f(x) = 2x + 1, \qquad g(x) = x^2 - 2x + 3.$$

(a) Find $f(-3)$ and $g(-3)$.

(b) Find and simplify equations defining the functions $f \circ g$ and $g \circ f$.

(c) Show that f is one-to-one and that g is not one-to-one.

(d) Show that f is a function from R onto R.

(e) If $h(x) = (x - 1)/2$, find equations defining the functions $f \circ h$ and $h \circ f$. How are f and h related?

6. Let f and g be functions from R into R, defined by

$$f(x) = x^2, \qquad g(x) = \frac{1}{1 + x^2}.$$

(a) If a and b are real numbers, find and simplify expressions for $f(a + b)$, $f(a) + f(b)$, and $f(2a)$.

(b) Determine the range of f.

(c) Show that the number $\frac{1}{2}$ is in the range of g.

(d) Find and simplify equations defining the functions $f \circ g$ and $g \circ f$.

7. **(a)** Let f and g be functions from N into N, where N is the set of natural numbers, defined by

$$f(x) = x^2, \qquad g(x) = x + 1.$$

Are f and g one-to-one? Why? Are f and g functions from N onto N? Why?

(b) Let F and G be functions from I into I, where I is the set of integers, defined by

$$F(x) = x^2, \qquad G(x) = x + 1.$$

Are F and G one-to-one? Why? Are F and G functions from I onto I? Why?

8. Give an example of a function from N onto B, where the set B is

    (a) the set of all natural numbers that are divisible by 3,
    (b) the set of all natural numbers greater than or equal to 100,
    (c) the set $\{0, 1\}$.

9. Let f be the function from R into I, defined for all real numbers x by

$$f(x) = \text{the greatest integer less than or equal to } x.$$

    (a) Find $f(\frac{10}{3})$, $f(-\frac{10}{3})$, $f(3)$, $f(-3)$.
    (b) Is f a mapping of R onto I? Explain.
    (c) Let a function g from N into N be defined by $g(n) = n - 2f(n/2)$. Show that $g(n) = 0$ if and only if n is even.
    (d) What is the range of the function g defined in part **c**? Explain.

10. Let $A = \{1, 2, 3, 4\}$ and $B = \{5, 6, 7, 8\}$. Consider the following four binary relations from A into B.

$$f = \{(1, 6), (2, 8), (3, 6), (4, 7)\},$$
$$g = \{(1, 8), (2, 7), (3, 6), (2, 5)\},$$
$$F = \{(1, 8), (2, 6), (3, 6)\},$$
$$G = \{(1, 8), (2, 6), (3, 7), (4, 5)\}.$$

    (a) Determine the domain and range of each of the four relations.
    (b) Which of the four relations is a function from A into B? Explain.
    (c) For those relations that are functions, which are one-to-one? Which are functions from A onto B? Explain.

11. Let $A = \{1, 2, 3, 4\}$ and let B denote the five-element set $B = \{a, b, c, d, e\}$. We define three functions f, g, h by means of the following tables, where f is a function from A into B and g and h are functions from B into A.

x	$f(x)$		x	$g(x)$		x	$h(x)$
1	a		a	2		a	2
2	b		b	3		b	3
3	c		c	1		c	4
4	d		d	1		d	—
			e	4		e	—

The fact that the elements d and e are followed by dashes in the h-table means that they are not in the domain of h.

(a) Which of the three functions are one-to-one? Explain.

(b) For those functions that are one-to-one, construct tables for the inverse functions.

(c) Construct tables for the functions $f \circ g$ and $g \circ f$, provided these functions are defined. If they are not defined, explain why.

(d) Do the same as in part **c** for the functions $f \circ h$ and $h \circ f$.

12. Which of the following sets are closed relative to the indicated operations? Justify your answers.

(a) The set of all odd integers; addition, multiplication.

(b) The set of all even integers; addition, multiplication.

(c) The set of all real numbers of the form $a + b\sqrt{2}$, where a and b are rational numbers; addition, multiplication.

(d) The set of all nonzero rational numbers; addition, multiplication.

(e) The family of all finite subsets of N; union, complementation.

(f) The family of all infinite subsets of N; union, intersection.

13. Let $A = \{a, b, c, d, e\}$ be a five-element set. Let g be the binary relation on A represented by the 0-1 matrix

$$
\begin{array}{c c}
 & \begin{array}{c c c c c} a & b & c & d & e \end{array} \\
\begin{array}{c} a \\ b \\ c \\ d \\ e \end{array} &
\left(\begin{array}{c c c c c}
0 & 1 & 0 & 0 & 0 \\
0 & 0 & 1 & 0 & 0 \\
0 & 0 & 0 & 1 & 0 \\
0 & 0 & 0 & 0 & 1 \\
1 & 0 & 0 & 0 & 0
\end{array}\right).
\end{array}
$$

(a) Show that g is a one-to-one function from A onto A.

(b) Determine the 0-1 matrix that represents the inverse of g.

(c) Determine the 0-1 matrix that represents the function $g \circ g$.

14. Let R be a binary relation defined on a finite set A, and let M be the 0-1 matrix that represents R.

(a) What can be said about M if R is a function from A into A? Explain.

(b) What can be said about M if R is a one-to-one function from A into A? Explain.

(c) What can be said about M if R is a function from A onto A? Explain.

15. Let $A = \{a, b, c\}$ and $B = \{0, 1\}$.

(a) Determine eight different functions with domain A and range in B. Describe the functions pictorially as in Figure 2.6-2.

(b) Determine nine different functions with domain B and range in A. Describe the functions as you did in part **a**.

16. Let f be a function from A into B. Let E be the binary relation defined on A by the rule

$$x \, E \, y \quad \text{if and only if} \quad f(x) = f(y).$$

(a) Show that E is an equivalence relation on A.

(b) Describe the E-equivalence classes for the specific example in which both A and B are the set I of all integers and f is defined by the equation $f(x) = x^2$.

17. Let A and B be subsets of S as defined below:

$$S = \{a, b, c, d, r, s, t\}, \quad A = \{a, b, r, s\}, \quad B = \{b, s, t\}.$$

Complete the following table and then verify that the four properties of characteristic functions listed in Example 3 are valid for this choice of sets.

x	$I_A(x)$	$I_B(x)$	$I_{A'}(x)$	$I_{A \cap B}(x)$	$I_{A \cup B}(x)$
a	1	0			
b	1	1			
c	0	0			
d					
r					
s					
t					

18. Use the sets given in Exercise 17 and an analogous table of values to illustrate each of these additional assertions about characteristic functions.

(a) $I_{(A \cup B)'}(x) = I_{A'}(x) I_{B'}(x).$

(b) $I_{(A \cap B)'}(x) = I_{A'}(x) + I_{B'}(x) - I_{(A' \cup B')}(x).$

A RECAPITULATION

2.7 Mathematical Structures

Chapter 1 and Sections 1 through 6 of this chapter presented fundamental concepts, terms, symbols, and methods used throughout mathematical studies. Chapters 3 through 5 present topics drawn from two major areas of discrete mathematics: graph theory and combinatorics. Before we make a transition from the general to the particular, it will be useful to recognize that much of what we have done can be reformulated in terms of the notion of a *mathematical structure,* a unifying concept that provides deeper insight into the nature and spirit of contemporary mathematics.

Starting from an intuitive concept of sets in Chapter 1, we learned how to compute algebraically with sets and to use set theory to define precisely the general concept of a binary relation from one set to another set. We saw that the important types of binary relations that can be defined from a set to itself are equivalence relations, partial order relations, and total order relations and that each type is distinguished by particular properties. Furthermore, we defined a function as another special type of binary relation and a binary operation as a special type of function. Thus we learned that sets, relations, operations, and functions are closely integrated concepts that together comprise the building blocks of a mathematical structure. Before we define the term, mathematical structure, we will recall some examples discussed in earlier sections.

In addition to learning about the algebra of sets, we learned about a similar algebra of statements or propositions, which was used in turn to describe the forms of direct and indirect proofs. We briefly studied the system of natural numbers, including the well-ordering property and mathematical induction. We were then introduced to graphs, 0-1 matrices, and matrix calculations, all in connection with binary relations. We even learned a few principles of counting that will be extended considerably in Chapter 3, and we will consider properties and uses of graphs in Chapters 4 and 5.

EXAMPLES

1. A *directed graph* $\mathscr{D}$ was defined in Section 2.2 as consisting of a nonempty set V together with a subset E of $V \times V$. When we recall that a subset of $V \times V$ is simply a binary relation r on V, we can restate the definition of a directed graph as a nonempty set V together with a binary relation r on V. This can be written symbolically as $\mathscr{D} = \{V; r\}$, where we assume that V is a nonempty set and that r is a binary relation on V. Such assumptions are called *axioms* or *postulates* of the structure $\mathscr{D}$.

2. A *graph* $\mathscr{G}$ was defined in Section 2.2 as consisting of a nonempty set V together with a subset E of $V \times V$ such that $(i, j) \in E$ if and only if $(j, i) \in V$. Hence we can recognize E as a symmetric binary relation s on V, so we write $\mathscr{G} = \{V; s\}$, and list the axioms of the structure $\mathscr{G}$ to be that V is nonempty and that s is a symmetric binary relation on V.

3. A *partially ordered set* was defined in Section 2.5 as consisting of a nonempty set S on which is defined a binary relation P that is either a strong partial order relation or a weak partial order relation. In the

former case the axioms are that P is irreflexive, asymmetric, and transitive; in the latter case the axioms are that P is reflexive, antisymmetric, and transitive. When viewed as a structure, a partially ordered set is thought of as an ordered pair $\mathscr{S} = \{S; P\}$, just as are the structures in Examples 1 and 2. Also in Section 2.5 a *totally ordered set* T was defined as a partially ordered set $\mathscr{T} = \{S; P\}$ in which P satisfies one additional axiom.   ∎

All the previous examples of mathematical structures consist of one nonempty set and one binary relation defined on that set. They differ from one another only to the extent that the axiom systems differ, and they are listed in the order of increasing complexity of the set of axioms. We now consider the natural number system $\mathscr{N}$ that has a more complex structure, even though it is quite familiar to us. To describe $\mathscr{N}$ as a structure, a natural approach might be to start by writing something like $\mathscr{N} = \{N; \leq; +, \cdot\}$ and then listing whatever axioms might be needed to capture precisely the properties of the natural numbers, including those of addition, multiplication, and ordering. The list would be fairly long, but it could be done. Instead, we present a modified form of the first axiomatic description of the natural numbers, presented nearly a century ago by the Italian mathematician Guiseppe Peano. Inasmuch as mathematicians in the twentieth and late nineteenth centuries have employed axiomatic methods so effectively, one wonders why scholars needed more than 2000 years to adopt the axiomatic approach of Greek geometers in studying other aspects of mathematics.

In Example 4, an axiomatic description of the natural number system, N denotes a nonempty set satisfying certain postulates, departing from our use of N to denote the set of natural numbers. Using N in two senses is justified because the objects of the set in this example have the usual properties of natural numbers.

4. The system $\mathscr{N}$ of natural numbers is a mathematical structure $\mathscr{N} = \{N; s\}$, where N is a nonempty set and s is a function from N into N such that the following postulates are satisfied.

   1. There is an element of N denoted by 0.
   2. For each x in N there is a uniquely determined element $s(x)$ in N.
   3. For each x in N, $s(x)$ is not 0.
   4. For each x, y in N, if $s(x) = s(y)$ then $x = y$.
   5. If M is any subset of N such that
      (a) 0 is in M, and
      (b) $s(y)$ is in M whenever y is in M, then $M = N$.

Postulate 1 states that N is not empty. Postulate 2 declares that s is a function from N to N; postulate 3 says that s is not onto N because 0 is not in its range, and postulate 4 assures us that s is a one-to-one function. The form of postulate 5 is reminiscent of the principle of induction.

From this elegant and ingenious set of axioms, Peano was able to *define* addition, multiplication, and the ordering of natural numbers and to prove all of their known properties. As you might expect, that task would divert us too far from our objective, so we add only the interpretation that s is called the *successor* function that assigns to each $x \in N$ its immediate successor $s(x)$, which we might expect to become known as "$x + 1$" after both 1 and $+$ have been defined from the axioms.

5. Even though you might not be familiar with this example, you should work through it carefully because this structure has important uses in combinatorics. In what follows we use the letter t as a formal symbol with no meaning attached to it. It is used only to create other formal symbols of the form

$$a_0 + a_1 t + a_2 t^2 + \cdots + a_n t^n + \cdots,$$

where a_i is an integer for $i = 0, 1, 2, \ldots$. Also the plus signs and exponents on t are to be interpreted as formal symbols and not as addition or multiplication of certain objects. Any symbol of that form is called a *formal power series* in the indeterminate t.

We now create a mathematical structure by describing how to "add" and to "multiply" formal power series. Let p and q denote formal power series, where

$$p = a_0 + a_1 t + a_2 t^2 + \cdots + a_n t^n + \cdots,$$
$$q = b_0 + b_1 t + b_2 t^2 + \cdots + b_n t^n + \cdots.$$

By definition the sum of p and q, written $p \oplus q$, is

$$p \oplus q = (a_0 + b_0) + (a_1 + b_1)t + (a_2 + b_2)t^2 + \cdots$$
$$+ (a_n + b_n)t^n + \cdots;$$

the product, written $p \odot q$, is

$$p \odot q = a_0 b_0 + (a_0 b_1 + a_1 b_0)t + (a_0 b_2 + a_1 b_1 + a_2 b_0)t^2 + \cdots$$
$$+ (a_0 b_n + a_1 b_{n-1} + \cdots + a_i b_{n-i} + \cdots + a_0 b_n)t^n + \cdots.$$

The sums and products formed to calculate the coefficients of powers of t in $p \odot q$ denote ordinary addition and multiplication of integers. Another way of stating the definition of $p \odot q$ is to say that for each nonnegative integer n, the coefficient of t^n in the product of p and q is the sum of all products of the form $a_r b_s$, where r and s range over all nonnegative integers such that $r + s = n$.

Observe that the coefficients of powers of t in $p \oplus q$ and $p \odot q$ are integers, because the set of integers is closed under ordinary addition and multiplication. If we let P denote the set of all formal power series in t, then this means that the set P is closed under the operations $\oplus$ and $\odot$. The resulting structure $\mathcal{P} = \{P; \oplus, \odot\}$ can be shown to have the following properties.

1. $\oplus$ and $\odot$ are commutative operations.
2. $\oplus$ and $\odot$ are associative operations.
3. $\odot$ is distributive over $\oplus$.
4. There exists an element z in P such that $p \oplus z = p$ for every p in P.
5. For each q in P there exists an element r in P such that $q \oplus r = z$, where z is the element described in 4.
6. There exists an element u in P such that $p \odot u = p$ for every p in P.

It is interesting to note that each property of the formal power series is also valid within the system $\mathcal{I} = \{I; +, \cdot\}$ of all integers. Any such structure is called a *commutative ring with unity*.

6. A *boolean algebra* is a mathematical structure $\mathcal{B} = \{B; \cup, \cap, '\}$, where B is any nonempty set, $\cup$ and $\cap$ are binary operations on B, and $'$ is a function from B onto B, and the following postulates are satisfied.

1. $\cup$ and $\cap$ are associative operations.
2. $\cup$ and $\cap$ are commutative operations.
3. $\cup$ is distributive over $\cap$, and $\cap$ is distributive over $\cup$.
4. There exist distinct elements z and u in B such that $x \cup z = x$ and $x \cap u = x$ for each x in B.
5. For each x in B there exists x' in B such that $x \cup x' = u$ and $x \cap x' = z$.

The element z, whose existence is assumed in Axiom 4, is called an *identity element* relative to the operation $\cup$ because $x \cup z = x$ for every x in B. Similarly, the element u is called an identity element relative to the operation $\cap$. When we apply the same language to the system of natural numbers, we say that 0 is an identity element relative to addition and 1 is an identity element relative to multiplication.

The most familiar example of a boolean algebra is the algebra of all subsets of a given set; the algebra of statements that we studied in Sections 1.6 – 1.8 is another familiar example.

We now want to use these few examples of mathematical structures as models from which to formulate a general description that is broad enough to encompass the mathematical systems discussed in this book, and indeed many others. To do so, we first look for common aspects of all these structures. Immediately we notice that each example has a nonempty set S of elements, even though the nature of those elements varies from example to example. Also each example has one or more ways of comparing or combining or otherwise manipulating those elements within the structure. In some examples there was one binary relation defined on S, and in another there were two binary operations and one binary relation. In Peano's description of $\mathcal{N}$ there was only a successor function defined from S into S; such a function is called a *unary operation on S*. It will suffice to say that the structure has a set R of binary relations on S and a set O of operations defined on S. Either of the sets R or O might be empty. Finally, the structure will be governed by a nonempty set P of postulates (or axioms or assumptions) about the elements of the sets S, R, and O.

> **Definition** A *mathematical structure $\mathcal{S}$* is a collection of four sets $\mathcal{S} = \{S; R; O; P\}$ such that
>
> **1.** S is a nonempty set.
> **2.** R is a set of relations on S.
> **3.** O is a set of operations on S.
> **4.** P is a nonempty set of postulates concerning S, R, and O.

For simplicity we usually modify the notation in the definition of a mathematical structure by listing the postulates separately, omitting the symbol P within the braces. Further, we often list the symbol for each relation in R and the symbol for each operation in O within the braces in place of the two symbols R and O. When no confusion is likely to arise, we sometimes use the same symbol for the structure itself and the set of elements of the structure; that is, given a boolean algebra $\mathcal{B} = \{B; \cup, \cap, '\}$, we might refer to that boolean algebra as B rather than $\mathcal{B}$.

Starting from such a bare bones structure, mathematicians prove theorems, assuming only the listed postulates. Because these theorems are deduced from the postulates alone, they will be valid for all specific examples of that structure, which is one of the efficiencies cited for the axiomatic method. As a body of theorems is developed for a given structure, new terms are defined to simplify the language. Thereby a set D of definitions and a set T of theorems also become evolving features of that mathematical structure.

But how do mathematicians choose the bare bones that form the skeleton of such a structure, and how do they choose suitable axioms that will endow the structure with useful properties? Definitive answers to such questions necessarily involve a discussion of artistic and intellectual creativity. We will not attempt to explore such questions here, but we should recognize that the structures mathematicians study are not the result of random selection. Rather, they evolve from mathematicians having numerous encounters and deep engagement with such structures in many concrete settings. For example, the structure known as a *group* (see Exercise 5) was encountered again and again from the end of the eighteenth century in many specific problems. However, it was not until quite late in the nineteenth century that the structure known as an *abstract group* was introduced. This enabled mathematicians to build an axiomatic theory, known as *group theory,* that encompassed many related but distinct problems, thereby unifying related studies and clarifying distinctions between them.

Exercises 2.7

1. Consider the mathematical structure $\mathcal{M} = \{M; \blacktriangleleft\}$, where M denotes the set of all 2-by-2 matrices having integer entries, and where $\blacktriangleleft$ denotes the binary relation on M defined by

   $A \blacktriangleleft B$ if and only if   the entry in row i and column j of A is less than or equal to the entry in row i and column j of B for all values of i and j.

   For example,

   $$\begin{pmatrix} 1 & 5 \\ 0 & 3 \end{pmatrix} \blacktriangleleft \begin{pmatrix} 4 & 5 \\ 2 & 4 \end{pmatrix}$$

   because $1 \le 4$, $5 \le 5$, $0 \le 2$, and $3 \le 4$.

   (a) Is the mathematical structure $\mathcal{M}$ a partially ordered set? Explain.
   (b) Is $\mathcal{M}$ a totally ordered set? Explain.

2. A *semigroup* is a mathematical structure of the form $\mathcal{S} = \{S; *\}$, where S is a nonempty set, $*$ is a binary operation on S, and $*$ is associative. Let M denote the set of all 2-by-2 matrices with integer entries, and let $*$ denote matrix multiplication. Verify that $\{M; *\}$ is a semigroup.

3. Let C be the set of all points inside and on the boundary of a given circle. Let $\mathcal{C} = \{C; \blacktriangleright; *\}$, where $\blacktriangleright$ is the relation defined by

$p \blacktriangleright q$ if and only if   the distance from p to the center of the circle is less than or equal to the distance from q to the center,

and $*$ is the operation defined by

$$p * q = \text{the midpoint of the segment from } p \text{ to } q.$$

(a) Show that the set C is closed relative to $*$.
(b) Is $*$ commutative? Explain.
(c) Is $*$ associative? Explain.
(d) Is $\blacktriangleright$ a partial ordering of C? Explain.

4. Let M denote the set of all 2-by-2 matrices having integers as entries, and let $+$ denote matrix addition.

(a) Show that $\{M; +\}$ is a semigroup as defined in Exercise 2.

A *monoid* is a semigroup $\{S; *\}$ such that S contains an element i having the property that for all x in S

$$x * i = x = i * x.$$

Any element having this property is called an *identity element* of S relative to the operation $*$. (There can be only one identity element in a monoid.)

(b) Show that $\{M; +\}$ is a monoid.

A *group* is a monoid $\{S; *\}$ with the property that for each x in S there exists an element x' in S such that

$$x * x' = i = x' * x,$$

where i is the identity of S. It follows that x' is uniquely determined by x and is called the inverse of x.

(c) Show that $\{M; +\}$ is a group.

5. Given a set G that is closed relative to a binary operation $*$, the mathematical structure $\mathcal{G} = \{G; *\}$ is called a *group* if and only if the following postulates are satisfied.

1. The operation $*$ is associative.

2. There exists an identity element i in G such that

$$y * i = i * y = y \qquad \text{for every element } y \text{ in } G.$$

3. For each element y in G there exists an element y' in G such that

$$y * y' = i = y' * y,$$

where i is the identity element described in postulate 2.

For each of the following structures determine whether that structure is a group, and support each answer with sufficient evidence.

(a) $\{N; +\}$ (the natural numbers with addition as the operation)
(b) $\{I; +\}$ (the integers with addition as the operation)
(c) $\{I; \cdot\}$ (the integers with multiplication as the operation)

6. Let R be the set of all real numbers, and let S be the set whose elements are the six functions f, g, h, F, G, H from R into R, defined by the equations

$$f(x) = x, \qquad g(x) = \frac{1}{1-x}, \qquad h(x) = \frac{1}{x},$$

$$F(x) = 1 - x, \qquad G(x) = \frac{x-1}{x}, \qquad H(x) = \frac{x}{x-1}.$$

(a) Show that S is closed under the operation $\circ$, where $\circ$ denotes function composition as defined in Section 2.6.
(b) Show that there is a function u in S such that $w \circ u = u \circ w = w$ for every w in S.
(c) For each w in S show that there exists a function w' in S such that $w \circ w' = w' \circ w = u$, when u is the function found in **b**.
(d) What additional fact must be verified to conclude that the structure $\{S; \circ\}$ is a group? (The structure is a group, but you are not asked to make that verification.)

7. Let $\odot$ denote the product operation defined for formal power series in Example 5, and let $p = 1 + t + t^2 + t^3 + \cdots + t^n + \cdots$. For each power series q calculate the terms up to and including the term involving t^5 of the sum $p \oplus q$ and of the product $p \odot q$.

(a) $q = p = 1 + t + t^2 + t^3 + \cdots + t^n + \cdots$
(b) $q = 1 + t + 2t^2 + 3t^3 + \cdots + nt^n + \cdots$
(c) $q = 1 - t + t^2 - t^3 + \cdots + (-1)^n t^n + \cdots$

8. As defined in Example 5, a formal power series may have terms with zero as coefficient. In that case we simply omit those terms. For example, the power series in which the coefficient of each even power of t is zero is written as $t + t^3 + t^5 + \cdots$, and any power series for which the coefficient of t^n is zero whenever $n > 2$ can be written in the form $a_0 + a_1 t + a_2 t^2$.

(a) Calculate the coefficients of t^9 and t^{15} in the product

$$(1 + t^3 + t^6 + t^9 + \cdots) \odot (1 + t^2 + t^4 + t^6 + \cdots).$$

(b) Calculate all terms up to and including the term involving t^5 in the product

$$(1 - t) \odot (1 + t + t^2 + t^3 + \cdots).$$

(c) Determine all terms of the product given in part **b**. Explain your answer.

Chapter 3

COMBINATORIAL MATHEMATICS

This chapter applies and extends the fundamental techniques of enumeration introduced in Sections 1.4 and 1.5 and builds upon the three principles presented there:

1. The Additive Rule of Or
2. The Multiplicative Rule of And
3. The Inclusion-Exclusion Principle

You should now review those principles.

Determining the size of a finite set is a problem belonging to the branch of contemporary mathematics called *combinatorics*. For the past three centuries the adjective form of that word, *combinatorial*, has been used with various mathematical topics, including algebra, number theory, topology, set theory, and applied mathematics. Currently, *Mathematical Reviews* lists combinatorics as a separate heading between set theory and number theory.

No matter how combinatorics is classified within mathematics, it is developing very rapidly and using techniques from many different parts of mathematics. Sections 3.1 through 3.7 discuss elementary techniques of enumeration; more advanced combinatorial techniques are introduced in Sections 3.8 through 3.10.

BASIC METHODS OF COUNTING

3.1 Combinatorial Problems

The members of a finite set can be counted in many ways. If the set is small, we can simply list the members and number the list with positive integers, starting with 1 and proceeding in the natural order until we reach the end of the list. But for a large set, that method is slow and error-prone. Consequently, more sophisticated strategies and techniques have been developed for counting large or intricately defined sets. Successful counting depends largely on one's skill and insight in selecting an appropriate method for the problem at hand. Such skill can best be developed by becoming thoroughly familiar with various methods of counting and by acquiring experience in applying these counting methods to a wide range of specific problems. This section identifies and illustrates some useful strategies and principles of counting.

First we emphasize the importance of *generalization* and *experimentation.* For example, to determine the number of 11-member committees that can be formed from the 100 members of the United States Senate, we should recognize that this question is a special instance of the following more general question:

How many different subsets of size k can be selected from the members of a set having n distinct members?

We might solve this problem by using some small values of k and n with $k \le n$. Often the reasoning used in solving the smaller examples can be used to solve the general problem. When we use this strategy, we hope to gain insight and understanding by investigating small versions of the same type of problem.

One very common method of counting the elements of a set is that of determining a sequence of well-defined steps for constructing all members of the set but no other objects. The number of objects constructed can then be calculated from the numbers of outcomes of the various steps in the construction process. We used this method in Section 1.4 to count the number of choices for two customers at Joe's Diner. The following example provides another illustration of counting by *construction.*

EXAMPLE 1   How many three-digit decimal numbers have three different digits, one of which is 6?

We can construct any such number by filling each of three adjacent positions with three different digits, where one digit is 6 and the left-hand digit is not 0. If 6 is placed in the left-hand position, then any of nine different digits can fill the middle position, and any of eight digits can fill the right-hand position. Thus, $(9)(8) = 72$ such numbers begin with 6. If 6 is placed in the middle position, eight nonzero digits are available for the left-hand position and eight remaining digits are available for the right-hand position. Hence, $(8)(8) = 64$ numbers can have 6 as the middle digit. A moment's reflection should convince us that there also are 64 numbers having 6 as the right-hand digit. Thus, the total count of numbers of the prescribed type is $72 + 64 + 64 = 200$. ■

Another general approach to counting uses *analogy and matching* by recognizing that the set to be counted has the same number of elements as some other set that is easier to count.

EXAMPLE 2   How many matches need to be scheduled to determine the champion of a tennis tournament in which there are 83 entrants?

Our first inclination might be to draw a large chart showing a possible pairing of players, where the winner of each match survives to play in the next round. But if we recall that each match determines one loser and that there is only one champion at the end of the tournament (and hence exactly 82 losers), we can deduce that exactly 82 matches, one less than the number of entrants, are needed. ■

The previous example also illustrates the technique for counting a *complement* of a set. The set of nonchampions in a tennis tournament is the complement of the set of champions within the set of entrants.

Induction and the related concept of *recursion* also provide means of counting. The Tower of Hanoi problem (Example 3 in Section 1.10) can be solved by induction: One move is needed for the 1-ring puzzle, three moves for the 2-ring puzzle, seven moves for the 3-ring puzzle, and we *guess* that $2^n - 1$ moves will suffice for the n-ring puzzle. Indeed, we can solve the n-ring puzzle by initially solving the $(n - 1)$-ring puzzle in $2^{n-1} - 1$ moves, using the top $n - 1$ rings; we then use one move to transfer the largest ring to the empty post, and we repeat the sequence of $2^{n-1} - 1$ moves needed to place the other $n - 1$ rings on the largest ring. The total number of moves is

$$(2^{n-1} - 1) + 1 + (2^{n-1} - 1) = 2(2^{n-1}) + 1 - 2 = 2^n - 1.$$

Induction can also be used to define numerical-valued functions on the set N of natural numbers, as illustrated in Example 3.

EXAMPLE 3   The *factorial* symbol ! can be defined inductively for every natural number n by the two statements

    1. $0! = 1$,
    2. $k! = k(k - 1)!$ for each $k > 0$.

Statement 1 is the *starting step* of the definition, and statement 2 is the *inductive step*. Mathematical induction then can be used to prove that

    3. $n!$ is defined for each natural number n.   ∎

The value of $n!$ increases rapidly as n increases, as shown in Table 3.1-1.

A *recursion equation* expresses the value $F(n)$ of a counting function F at a natural number n in terms of one or more values of F at natural numbers less than n. In the Tower of Hanoi example the recursion equation can be written as

$$H(n) = 2H(n - 1) + 1 \quad \text{when } n > 1, \quad \text{and} \quad H(1) = 1.$$

Another example obtained from the definition of the Fibonacci sequence given in Exercise 12 of Section 1.10 is

$$F(1) = 1 = F(2),$$
$$F(k) = F(k - 1) + F(k - 2) \quad \text{for all } k > 2.$$

Recursion equations often occur in combinatorics, and recursive computation is an important aspect of computer science. After this next example we defer further examination of recursion until Section 3.8.

n	$n!$	n	$n!$
0	1	5	120
1	1	6	720
2	2	7	5,040
3	6	8	40,320
4	24	9	362,880

Table 3.1-1

EXAMPLE 4   A seven-foot basketball player finds it difficult to climb stairs one at a time and is more comfortable taking either 2 or 3 stairs in one stride. In how many ways can a tall person climb a 16-stair staircase by taking 2 or 3 stairs at each step?

For $k > 1$, let $w(k)$ denote the number of ways that a k-stair staircase can be climbed in the manner described. Then we have $w(2) = 1$, $w(3) = 1$, $w(4) = 1$. To climb k stairs 2 or 3 at a time, the climber must first reach either stair $k - 2$ or stair $k - 3$, and from each of those stairs there is only one way to finish. Hence

$$w(k) = w(k - 2) + w(k - 3) \qquad \text{for } k > 4.$$

Thus, we have, for example,

$$w(5) = w(3) + w(2) = 1 + 1 = 2,$$
$$w(6) = w(4) + w(3) = 1 + 1 = 2,$$
$$w(7) = w(5) + w(4) = 2 + 1 = 3,$$
$$w(8) = w(6) + w(5) = 2 + 2 = 4.$$

Continuing in this way, we find the first few terms of the sequence to be

$$1, 1, 1, 2, 2, 3, 4, 5, 7, 9, 12, 16, 21, 28, 37, \ldots,$$

so the correct answer is $w(16) = 37$. If we had wanted to know the answer for a 116-stair staircase, we would want to have a less tedious method of evaluating $w(116)$, such as a formula for $w(n)$ or a computer program to calculate $w(n)$ recursively. ∎

Another principle of counting seems almost too obvious to mention, but its applications are sometimes quite subtle.

The Pigeonhole Principle   If n pigeons roost in k pigeonholes ($n > k$), then at least one pigeonhole must contain more than one pigeon.

This principle, for example, allows us to deduce that two or more residents of Detroit have the same number of hairs on their heads, because the number of residents (pigeons) is much larger than the maximum number of hairs (pigeonholes) that can grow on any human head.

Now suppose that n pigeons roost in k pigeonholes and that $n > 2k$; then at least one pigeonhole must contain more than two pigeons. To prove that assertion, we use an indirect proof. Suppose that all n of the pigeons are roosting, but that none of the k pigeonholes contains more than two pigeons. Then the number of roosting pigeons does not exceed $2k$, and $2k < n$, contradicting the assumption that all pigeons are roosting. Using essentially the same argument, we can prove the following more general statement of the pigeonhole principle.

> **The Generalized Pigeonhole Principle** If n pigeons roost in k pigeonholes, and if $n = mk + r$, where m and r are positive integers with $r < k$, then at least one pigeonhole contains more than m pigeons.

This form of the pigeonhole principle is easy to rephrase in terms of arithmetic means.

The number of pigeons in some pigeonhole is at least average.

To see this, divide each side of the equation in the generalized principle by k to obtain

$$\frac{n}{k} = m + \frac{r}{k}.$$

The number n/k is the average (arithmetic mean) number of pigeons in each pigeonhole, and it is larger than the integer m by a positive fraction, r/k. But each pigeonhole contains an integral number of pigeons, and if each pigeonhole were to contain m or fewer pigeons, then at most mk pigeons are roosting, contrary to the equation

$$n = mk + r \qquad \text{with } 0 < r < k.$$

So at least one pigeonhole must contain $m + 1$ or more pigeons, as claimed.

EXAMPLE 5 From the set $A = \{1, 2, 3, \ldots, 58\}$ of the first 58 positive integers, choose any subset B of 30 numbers. Prove that B must contain 2 numbers a and b such that b is divisible by a.

Observe that B contains more than half of the numbers in A, and that any positive integer x can be written uniquely in the form

$$x = 2^p y,$$

where p is a nonnegative integer and y is odd ($1 \leq y \leq x$). By writing each number of B in this form, we obtain 30 values of y, each of which is in A. Because A contains only 29 odd integers, the pigeonhole principle guarantees that 2 of those values of y are equal. That is, 2 distinct numbers a and b of A have the form $a = 2^p y$, $b = 2^q y$ for some y in A and for some p and q with $p < q$. Then

$$b = 2^{q-p}(2^p y) = 2^{q-p}a,$$

which shows that b is divisible by a. ∎

Application of Counting in Probability Theory

Consider an experiment that can be carried out repeatedly and that results (each time it is performed) in exactly one outcome out of a finite set of outcomes. A set of all possible outcomes of a single trial of the experiment is called a *sample space S* for that experiment. Any subset E of S is called an *event,* and any one-element subset $\{x\}$ of S is called a *simple event.*

A *probability function p* on a sample space S is a function that assigns a number $p(x)$ to each element x of S in such a way that the following conditions hold:

1. For each x in S, $0 \leq p(x) \leq 1$.

2. The sum of all numbers $p(x)$ as x varies over S is exactly 1.

If S is a finite sample space with probability function p defined and E is any event (subset of S), then the *probability $p(E)$* of the event E is defined as the sum of all numbers $p(x)$ as x varies over E.

An important special case occurs when the probability function p assigns to each x in S the same number, which must then be given by the formula

$$p(x) = \frac{1}{n(S)} \qquad \text{for each } x \in S.$$

With this definition of p, S is called an *equiprobability space.* In that case the probability of any event E is

$$p(E) = \frac{n(E)}{n(S)}.$$

This definition assigns to each event E a probability that is proportional to the number of simple events in E. Thus, the calculation of the probability of any event E in a finite equiprobability space S is reduced to counting the number of elements of E and the number of elements of S.

EXAMPLE 6   Mr. X always wears one of three colors of socks: brown, gray, or tan. He doesn't bother to arrange his socks in pairs after they are laundered but dumps all of them in a drawer. When dressing, he draws socks randomly from the drawer, one at a time, until he obtains a pair. On a certain day there are 4 brown, 4 gray, and 4 tan socks in the drawer. Determine the probability that he first obtains a pair of socks on the kth draw.

Consider the possible outcomes of this experiment to be an ordered sequence of the letters B, G, and T according to the order in which the colors appear on successive draws. Mr. X cannot obtain a pair in fewer than 2 draws, and he is certain to obtain a pair on or before the fourth draw, because by the pigeonhole principle we have 4 socks (pigeons) to assign to 3 colors (pigeonholes).

Therefore, we want to determine the probabilities for $k = 2$ and $k = 3$. The total number of 2-sock draws is

$$N(2) = 12(11) = 132$$

because there are 12 socks to draw from on the first draw, and 11 on the second draw. Thus each 2-sock draw has probability $\frac{1}{132}$. No matter which of 12 socks is drawn first, only 3 socks will complete the pair on the second draw; hence a pair can be drawn by 2 draws in

$$S(2) = 12(3)$$

ways. Thus the probability of obtaining a pair in 2 draws is

$$p(2) = \frac{12(3)}{12(11)} = \frac{3}{11}.$$

We can also calculate this probability directly by realizing that the outcome of the first draw is immaterial because any first draw leaves 3 socks of the chosen color and 8 socks of other colors.

As an exercise, calculate $p(3)$, which will complete this example because $p(1) = 0$ and $p(2) + p(3) + p(4) = 1$.   ∎

Exercises 3.1

1. Calculate the value of $F(13)$, the thirteenth Fibonacci number.

2. Let m and n denote natural numbers. Determine necessary and sufficient conditions for $m!/n!$ to be a natural number. Justify your answer.

3. How many positive integers less than 1000 have no repeated decimal digit?

4. Five different points are chosen within an equilateral triangle having each side of length 2 inches. Show that two of those points must be no more than 1 inch apart.

5. Let A be a 10-by-10 matrix where each entry is 0 or 1. If the sum of all entries is 72, then prove for some integers i and j that the sum of the entries in row i plus the sum of the elements in column j exceeds 15.

6. Let S be a set of distinct integers having the property that, for each pair a, b of distinct integers in S, neither $a + b$ nor $a - b$ is divisible by 10. Prove that S contains six or fewer integers.

7. Is it possible to arrange the numbers 1 through 12 around a circle in such a way that the sum of five consecutive numbers never exceeds 32? Explain.

8. Suppose that n lines are drawn on the plane in such a way that no two lines are parallel and no three lines intersect in a single point. Into how many regions is the plane separated by those lines, given that

   (a) $n = 0$, 1, 2, 3, 4?
   (b) $n = 999$?

   Explain your method.

9. If a single die is rolled six times, what is the probability of rolling either 2 or 3 at least once?

10. Determine the probability that an integer strictly between 26 and 92 is divisible by either 3 or 7 but not both.

11. Complete Example 6 by calculating $P(3)$ and $P(4)$.

12. An urn contains seven red balls, six white balls, and five blue balls.

    (a) What is the smallest number of balls that must be drawn to be certain that two balls of the same color are drawn?
    (b) What is the smallest number of balls that must be drawn to be certain that two white balls are drawn?
    (c) If only one ball is drawn, what is the probability that it is white?

(d) If exactly two balls are drawn, what is the probability that at least one is white?

13. Determine the minimum number of slices required to slice one 3-inch cube of cheese into twenty-seven 1-inch cubes, given that you may restack the pieces before making the second and third slices. Prove your answer.

14. Imagine a checkerboard (8-by-8 squares) from which 2 diagonally opposite corner squares have been removed, leaving a board with 62 squares. Can the 62 squares be covered by correctly arranging a set of 31 rectangles (each of dimension 2 squares by 1 square)? Prove your answer.

15. Let $x_1, x_2, \ldots, x_p$ be any sequence X of p distinct positive integers, where $p = n^2 + 1$ for some integer n. Any subsequence can be obtained by removing some, but not all, of the x_i without altering the relative positions of the remaining numbers. For example, 3, 8, 5, 26 is a subsequence of 9, 3, 2, 17, 8, 30, 5, 26, 14, 11. Furthermore, 30, 26, 14, 11 is a *decreasing* subsequence, whereas 9, 17, 26 is an *increasing* subsequence.

Prove that an arbitrary sequence X of $p = n^2 + 1$ distinct integers must have either a subsequence of length $n + 1$ that is decreasing or a subsequence of length $n + 1$ that is increasing.

3.2 Arrangements of Distinct Objects

PROBLEM   In how many different ways can five dice be stacked to form a vertical tower?   ■

Although we might not be prepared to solve this problem, we can learn much about combinatorics by thinking about the meaning of such a question. This kind of question becomes important when we try to choose a strategy for solving a puzzle or problem. If there are only a few ways to stack the dice, a reasonable (though uninspired) method would be to check each way, one at a time. But if the number of cases is sufficiently large, a case-by-case approach requires an unreasonable expenditure of time, and we should seek a more enlightened strategy.

One of the first issues that needs to be settled concerns the meaning of the word "different" in this problem. If the dice are indistinguishable from one another, do we care whether they are stacked in a particular order, say *abcde* rather than *daceb*? Are the faces of the dice numbered in a uniform way? For simplicity, suppose that the dice are alike, with the faces numbered in the standard way, in which the sum of the numbers on opposite faces is 7, and with the numbers distributed in such a way that when any die is placed with the 1-face down and the 2-face pointed north, then the 3-face is pointed east. Because we can choose any of the six faces to be the down face and then choose any of the four lateral faces to be the north face, we have $6(4) = 24$ ways of placing the first die relative to a given geographical direction. Using the Multiplicative Rule of And in this way for each of the five dice, we conclude that the tower of five dice can be constructed in $(24)^5$ ways relative to a given geographical direction. If geographical orientation is not a criterion for distinguishing between two towers, then the designation of "north" was arbitrary for the bottom cube, so we have overcounted by a factor of 4, and the number of different towers would be $6(24)^4 = 1,990,656$.

We are now ready to begin a systematic study of particular types of combinatorial problems that can be solved by elementary counting techniques. Sections 3.2 through 3.7 proceed from the simple to the more complicated examples centered on two basic categories of problems:

Arrangements (or permutations)

Selections (or combinations)

Each category can be subdivided into smaller classes according to special types of restrictions that might or might not occur in a given problem of that category. Moreover, a given combinatorial question might involve more than one of these basic forms of problems.

To count the number of *distinguishable* end results of arranging or selecting a set of objects, we need to have clearly in mind the sense in which two end results are regarded as distinguishable. This sense depends not only on whether we are counting sets (selections) or ordered sets (arrangements) but also on whether the objects themselves are distinguishable one from another.

A given set of n objects has exactly one of these three properties:

1. Each object is *indistinguishable* from every other object of the set. In this case the objects are said to be *identical.*

2. Each object is *distinguishable* from every other object of the set. Then the objects are said to be *distinct* or *different,* meaning that *no two* objects of the set are alike.

3. Neither property 1 nor property 2 holds. Then there would be m *types* of objects in the set, where $1 < m < n$, with k_i objects of type i for $i = 1, 2, \ldots, m$ and with $k_1 + k_2 + \cdots + k_m = n$, such that for each i all objects of type i are identical.

To simplify the language a bit, let us agree that when we count "the number of ways" we always mean "the number of distinguishable (or distinct or recognizably different) end results."

Another matter of language needs to be clarified. From the definition of equality of sets, it follows that the set {1, 1, 2, 2, 2} is equal to the set {1, 2}. Although a multiple listing of identical objects is ignored in set theory, it is essential in combinatorics. In Section 1.2 we introduced the term *multiset* to refer to an intended listing of elements with repetition. However, as another attempt to simplify language, we shall speak of a "set" (instead of a "multiset") of n identical objects. Usually what we refer to is an imagined collection of n physical objects among which we make no distinction for the purpose at hand. For example, we might speak of a set of 18 identical jellybeans if we are not interested in distinguishing large from small, round from ovate, green from yellow, licorice from cinnamon, and so on.

Definitions Let S be a set of n distinct objects, and let k be a positive integer ($k \leq n$).

1. An *arrangement* (or *permutation*) of S is an ordered n-tuple of all elements of S.
2. A *k-arrangement* (or *k-permutation*) of S is an ordered k-tuple of distinct elements of S.

Observe that an arrangement of S is the same as an n-arrangement of S, when S has exactly n distinct elements. We shall use the words "arrangement" and "permutation" interchangeably. When $k < n$, a k-permutation involves two operations, namely, selecting k of the n elements of S and then arranging them in some order along a line in the positions of a k-tuple or one in each of k pigeonholes or boxes that are positioned linearly.

Notation The symbol $P(n, k)$, where $0 \leq k \leq n$, denotes the number of k-permutations of a set of n distinct elements.

Theorem 3.2-1

$$P(n, k) = \frac{n!}{(n - k)!}.$$

Proof Recall the definition of $n!$ ("n factorial") given in Example 3 of Section 3.1. To evaluate $P(n, k)$ (the number of permutations of k

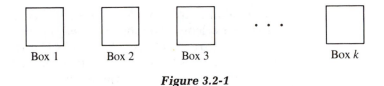

Figure 3.2-1

objects selected from a set of n distinct objects), we consider a row of k empty boxes, numbered from left to right as in Figure 3.2-1. We need to fill the first box *and* fill the second box *and* fill the third box, and so on. Box 1 can be filled in n ways. Then there are $n - 1$ objects available for filling box 2, so box 1 and box 2 can be filled in $n(n - 1)$ ways. Then box 3 can be filled in $n - 2$ ways, so box 1 and box 2 and box 3 can be filled in $n(n - 1)(n - 2)$ ways. We continue in this way; after box $k - 1$ has been filled, $k - 1$ objects have been put in boxes, so $n - (k - 1)$ objects remain as choices to fill box k. Hence, all k of the boxes can be filled in

$$n(n - 1)(n - 2) \cdots (n - k + 1) = P(n, k)$$

ways. This equation can be written more compactly as

$$P(n, k) = \frac{n!}{(n - k)!},$$

which is obtained by multiplying and dividing by $(n - k)!$ and by recognizing the revised numerator to be $n!$. ∎

EXAMPLE   The manager of a baseball team almost invariably assigns the pitcher to the last of the 9 positions in the batting order (when no designated hitter is used). Assume that a team roster has 8 pitchers, 3 catchers, and 12 other players who can play any of the other 7 positions. Also assume that the catcher will bat in the eighth position. How many different batting orders can the manager make?

Think of a list of 9 numbered blank spaces into which the manager writes the name of the player batting in that position for today's game. Slot 9 (the pitcher) can be filled in 8 ways. Slot 8 (the catcher) can be filled in 3 ways. The first 7 slots can be filled in $P(12, 7)$ ways. Hence, the manager has

$$8(3)P(12, 7) = \frac{24(12!)}{5!} = 95,800,320$$

ways to choose a batting order (which shows why managers so rarely select the best lineup)! ∎

It is sometimes appropriate to consider *circular* permutations instead of linear ones. How many different ways can a dinner party of eight persons be seated around a circular table (where there is no distinguishing position to regard as the head of the table or line)?

To count the number of ways in which k persons can be selected from n persons and be seated around a circular table, we pretend that there is a designated "head of the table," choose an arbitrary chair for that position, and regard the k chairs as being in a line moving clockwise around the table from the head. The k linearly ordered chairs moving clockwise from the head can be filled in $P(n, k)$ ways. But for each distinct circular order around the table, there are k copies of that circular order among the $P(n, k)$ linear orders, corresponding to each of the k persons sitting in the head position without changing the circular order. Hence, we calculate the number of circular orders by dividing $P(n, k)$ by k:

$$\frac{P(n, k)}{k} = \frac{n!}{(n-k)!k} = \frac{n(n-1) \cdots (n-k+1)}{k}.$$

The number $P(n, k)/k$ is an integer (that is, $P(n, k)$ is divisible by k) because among any k successive integers exactly one of them is divisible by k.

Exercises 3.2

1. The number $P(n, k)$ counts the number of distinct ways in which k objects can be selected from n different objects and arranged linearly, so we know that $P(n, k)$ must be a positive integer for each $k = 0, 1, \ldots, n$. Use the numerical ratio for $P(n, k)$ given in Theorem 3.2-1 to explain arithmetically why that ratio reduces to an integer.

2. To encourage the concept of friendly relations between two teams meeting for the state basketball championship, the members of each squad line up in any order before the game; then the squads file past each other so that each player shakes the hand of each opponent. One squad has 14 players, the other 12.

   (a) In how many different orders can the two squads form two lines?
   (b) For each way the squads line up, how many handshakes are exchanged? Make your reasoning clear in each case.

3. Use the formula for $P(n, n)$ to explain why 0! should be defined to have the value 1.

4. How many different five-letter words can be formed from the letters

of the word "variety"? (A "word" is any arrangement of one or more letters.) Show your reasoning.

5. Three couples are waiting in line at a restaurant for a table for six to become available.

   (a) In how many different orders can those six persons arrange themselves in line?

   (b) In how many different orders can they align themselves so that each is next to his or her partner?

   (c) In how many different orders can they align themselves so that no person is next to his or her partner?

6. Assume that the three couples in Exercise 5 are seated at a round table at which all places are identical.

   (a) In how many different arrangements can the six persons be seated?

   (b) In how many ways can they be seated so that each sits opposite his or her partner?

   (c) In how many ways can they be seated so that each sits next to his or her partner?

7. Fifteen different prizes are to be awarded by random selection to 15 children who attend a community picnic on July 4. If 213 children attend the picnic, in how many orders can the names of the winning children be drawn? (Do not attempt to calculate the numerical value of your answer.)

8. How many distinguishable charm bracelets can be made by hanging 15 different charms at equal intervals along a gold-link chain that has a closing clasp? (Take into account the positions of the charms relative to the clasp and the number of ways that a given bracelet can be placed around a wrist.)

9. How many different four-digit numbers have no two digits alike?

10. How many different license plates can be made to consist of four different digits, the first of which must exceed 1?

11. Four different persons are to be elected to the positions of president, vice-president, secretary, and treasurer of the Mathematics Club. In how many ways can the officers be elected if

    (a) there are 16 persons willing to serve in any office,

    (b) 12 persons are willing to serve only as vice-president, secretary, or treasurer, and 4 are willing to serve only as president?

12. Twelve different books (5 calculus books, 4 algebra books, and 3 statistics books) are to be arranged on a shelf. In how many ways can it be done if

     (a) the arrangement is unrestricted,
     (b) the statistics books must be on the left,
     (c) the books of each subject area must be kept together?

13. Prove that

$$(n + 1)! = 1 + \sum_{i=1}^{n} i(i!).$$

3.3   Selections from Distinct Objects

Section 3.2 described a k-arrangement of a set S of n distinct elements as the result of two successive operations: the selection of k elements from S, followed by a linear ordering of the selected elements. We now consider the number of ways in which k elements can be selected from n distinct elements.

Definition 3.3-1   A k-*selection* (or k-*combination*) from a set S of n distinct elements is a k-element subset of S.

Terminology in combinatorial mathematics is not standard; for example, the word "sample" is sometimes used as a synonym for "selection." The traditional terms for "arrangement" and "selection" are "permutation" and "combination," respectively, and this tradition is reflected in the notation.

Notation   The number of k-selections from an n-element set is denoted by $C(n, k)$.

Theorem 3.3-1

$$C(n, k) = \frac{n!}{(n - k)!k!}.$$

Proof   See Exercise 7. First justify the equation

$$C(n, k)P(k, k) = P(n, k)$$

by interpreting $P(n, k)$ as the number of different ways in which k objects can be first selected from a set of n distinct objects and then arranged along a line. Finally, use the Multiplicative Rule of And. ∎

The proof outlined for Theorem 3.3-1 is an example of a *combinatorial proof*. A combinatorial proof establishes a relationship between two or more combinatorial symbols by describing a set that can be counted by each such symbol and identifying a numerical relation between the sizes of those sets. Other examples of combinatorial proofs occur in Theorems 3.3-2, 3.3-3, and 3.3-4.

EXAMPLE 1 Determine the number of committees of five members that can be chosen from a list of seven students.

Each committee is a selection (a subset), so we need to evaluate $C(7, 5)$;

$$C(7, 5) = \frac{7!}{2!\,5!} = \frac{7(6)5!}{2!\,5!} = \frac{42}{2} = 21.$$

Thus, 21 committees of five students can be formed. More generally, any seven-element set has 21 different subsets of five elements each. ∎

It will be instructive to examine the values of $C(7, k)$ for all k from 0 to 7, shown in Table 3.3-1. In this form the symmetry of this sequence of numbers is unmistakable, and after a moment's reflection we can clearly see why:

Every selection of k elements from a set of n distinct elements leaves a selection of $n - k$ elements.

This reason provides a combinatorial proof of Theorem 3.3-2. If you prefer an algebraic argument, observe what happens to the calculation in Theorem 3.3-1 when k is replaced by $n - k$ on each side of the equation.

Theorem 3.3-2 $C(n, n - k) = C(n, k)$.

k	0	1	2	3	4	5	6	7
$C(7, k)$	1	7	21	35	35	21	7	1

Table 3.3-1

If we add all the values of $C(7, k)$ in Table 3.3-1, we are led to another property of $C(n, k)$. The total turns out to be 128, which can be written as 2^7.

Theorem 3.3-3

$$\sum_{k=0}^{n} C(n, k) = 2^n.$$

Proof The sum on the left side can be written

$$C(n, 0) + C(n, 1) + C(n, 2) + \cdots + C(n, n).$$

But $C(n, k)$ is the number of k-element subsets of an n-element set. If we sum all such numbers from 0 through n, we have counted all the subsets of an n-element set; the total is 2^n. (See Exercise 8 of Section 1.10.) ∎

Theorem 3.3-4

$$C(n + 1, k) = C(n, k) + C(n, k - 1) \qquad \text{for } 1 \le k \le n.$$

Proof See Exercise 9. ∎

This theorem allows us to compute all the values of $C(8, k)$ from the information in Table 3.3-1, the symmetry of $C(n, k)$, and the values $C(n, 0) = C(n, n) = 1$ for all n. Thus,

$$C(8, 0) = 1,$$
$$C(8, 1) = C(7, 1) + C(7, 0) = 7 + 1 = 8,$$
$$C(8, 2) = C(7, 2) + C(7, 1) = 21 + 7 = 28,$$
$$C(8, 3) = C(7, 3) + C(7, 2) = 35 + 21 = 56,$$
$$C(8, 4) = C(7, 4) + C(7, 3) = 35 + 35 = 70.$$

By now we recognize these properties of $C(n, k)$, and we recall that we used them in relation to the binomial theorem for expanding the expression $(x + y)^n$ for any natural number n. The connection is not at all hard to discover if we think about what $(x + y)^n$ means:

(∗) $(x + y)(x + y) \cdots (x + y),$

the product of n identical factors, $x + y$. Repeated use of the distributive law,

$$(a + b)c = ac + bc,$$

tells us that if we expand (*) without collecting terms we obtain the sum of 2^n terms, where each term is of the form $x^{n-p}y^p$ for integral values of p from 0 to n. When we collect all terms for a single value of p, say $p = t$, we obtain

$$c_t x^{n-t} y^t,$$

where c_t, the number of like terms that were collected, is a positive integer called a *binomial coefficient*. We claim that $c_t = C(n, t)$, the number of selections of t objects from a set of n objects. Looking back at (*), we realize that the binomial coefficient c_t simply counts the number of different ways that y can be chosen from exactly t terms of the form $x + y$ when n such terms are available. (Of course, x is then chosen from each of the remaining terms.)

Theorem 3.3-5 *(The Binomial Theorem)* Let x and y be numbers, and let n be a natural number. Then

$$(x + y)^n = \sum_{k=0}^{n} C(n, k)x^{n-k}y^k.$$

The binomial coefficients are sometimes presented in a tabular form, illustrated in Table 3.3-2, known as *Pascal's triangle,* in honor of the seventeenth-century French mathematician Blaise Pascal.

If we apply the binomial theorem with $x = 1$, we obtain a polynomial in y that can be written in the form

$$(1 + y)^n = \sum_{k=0}^{n} C(n, k)y^k$$
$$= C(n, 0)y^0 + C(n, 1)y^1 + C(n, 2)y^2 + \cdots + C(n, n)y^n.$$

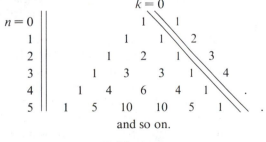

and so on.

Table 3.3-2

This expression is a polynomial function of degree n in y such that, for each k from 0 to n, the coefficient of y^k *automatically counts* the number of subsets of size k in a set of size n. Such a polynomial is a particular type of *generating function* for the problem of determining the number of selections of k elements from a set of n elements. We return to the topic of generating functions in Section 3.9.

EXAMPLE 2   How many different six-letter "words" can be formed from the letters of the word "tatters"?

   This question is a variation of our current discussion, because three of the seven given letters are identical. To form a six-letter word, we must fill six boxes so that each box has one letter. Either two or three of the boxes must contain t. If only two boxes contain t, then some set of four boxes must contain a, e, r, s in some order. We have $C(6, 4)$ ways to choose those four boxes and $P(4, 4)$ ways to assign letters to them. The other two boxes contain t, so there are $C(6, 4)P(4, 4)$ words with two t's. If three boxes contain t, then we can select three letters other than t in $C(4, 3)$ ways, choose three boxes out of six to contain them in $C(6, 3)$ ways, and distribute the letters other than t in the chosen boxes in $P(3, 3)$ ways. Each unchosen box contains t. Altogether the number of different six-letter words is

$$C(6, 4)P(4, 4) + C(4, 3)C(6, 3)P(3, 3) = 15(4!) + 4(20)3! = 840.$$

Note carefully the use of the Additive Rule of Or because of the two alternative frequencies of the letter t and, for each of those alternatives, the use of the Multiplicative Rule of And because of the successive steps required to form words. Note also that $C(6, 4)P(4, 4) = P(6, 4)$, that $C(4, 4) = 1$, and that $C(6, 3)P(3, 3) = P(6, 3)$. Therefore the calculation can be expressed in the symmetric form

$$C(4, 4)P(6, 4) + C(4, 3)P(6, 3).$$

Section 3.5 considers arrangements and selections that permit repetition of the objects, extending the considerations of this example.   ■

   Binomial coefficients are related to each other in many interesting ways, a few of which are given in the exercises for this section. For proving that a conjectured relationship is valid, we usually have a choice of methods. A direct algebraic verification is sometimes very easy, as in Theorem 3.3-2, for example. Sometimes induction is a good choice. A direct combinatorial argument, when one can be found, will show not only that a result is true but also why it is true. Proofs given for the first four theorems of this section are examples of combinatorial arguments; another is given in Example 3.

EXAMPLE 3 Prove the following equality for all $n \geq k \geq m$ in N:

$$C(n, k)C(k, m) = C(n, m)C(n - m, k - m).$$

The product on the left can be interpreted as the number of different ways to choose a work detail of k scouts out of a troop of n scouts, and then to choose m of those workers to sweep out the mess hall. The product on the right then is the number of different ways to select m scouts from the troop to sweep the mess hall and then to choose $k - m$ other workers from the depleted troop of $n - m$ scouts. Since each side counts the number of ways in which k scouts can be chosen from n scouts and assigned to two work groups of size m and $k - m$, the two products are equal. ∎

Exercises 3.3

1. A standard deck of cards for bridge, poker, and many other games has four suits (spades, hearts, diamonds, and clubs) and 13 cards in each suit (2 through 10, jack, queen, king, and ace).

    (a) How many different bridge hands are there? Write your answer in symbolic form (unevaluated) because it is larger than 635 billion. (A bridge hand is any unordered selection of 13 cards.)

    (b) What is the probability that a bridge hand contains four cards of one suit and three cards of each of the other three suits?

    (c) What is the probability that a bridge hand contains exactly two aces?

2. Write the horizontal lines in Pascal's triangle for $n = 6$, 7, and 8.

3. Two brothers inherited a library of 30 mathematics books, 30 history books, and 40 novels. One brother was more interested in mathematics than history, whereas the reverse was true for his brother. They decided that each would receive 20 novels, 25 books of his preferred subject, and 5 books of the remaining type. In how many ways can the books be divided according to this plan?

4. A poker hand consists of five cards from a standard deck. Write an expression for the probability of drawing

    (a) one pair (exactly two cards of the same denomination and one card of each of three other denominations),

    (b) two pairs (two cards of each of two different denominations and one card of a third denomination),

(c) triplets (exactly three cards of the same denomination and one each of two other denominations).

(d) Determine the relative ratios of those three probabilities.

5. Showing your method, determine which of the following products is larger: $C(20, 5)C(15, 11)$ or $C(19, 5)C(16, 12)$.

6. Twenty points are placed on a circle, and all chords joining pairs of those points are drawn. Assuming that no three chords intersect in a single point, how many points of intersection of those chords lie in the interior of the circle (not on the circumference)?

7. Prove Theorem 3.3-1.

8. Prove by algebraic computation that

(a) $C(n + 1, k) = \dfrac{(n + 1)C(n, k)}{n + 1 - k}$;

(b) $C(n, k + 1) = \dfrac{(n - k)C(n, k)}{k + 1}$;

(c) $C(n + 1, k + 1) = \dfrac{(n + 1)C(n, k)}{k + 1}$.

9. (a) Explain how the claim of Theorem 3.3-4 can be observed in Pascal's triangle.

(b) Prove Theorem 3.3-4 by algebraic verification.

(c) Prove Theorem 3.3-4 by a combinatorial argument.

10. If $1 \le k \le n$, consider the assertion that

$$C(n, k) = \sum_{m=k}^{n} C(m - 1, k - 1).$$

(a) Explain how this assertion can be observed in Pascal's triangle.

(b) Prove the assertion for an arbitrary natural number n by using induction on n.

11. Write in symbols the statement obtained by letting $x = 1$ and $y = 1$ in the binomial theorem. Then interpret in words the meaning of that statement. ("Interpret" means to do more than "translate.")

12. (a) Use Theorem 3.3-2 on each term in the sum of the formula in Exercise 10 and thus obtain a different form of expression for $C(n, k)$.

(b) Explain how the latter expression can be observed in Pascal's triangle.

13. Let $x = 1$ and $y = -1$ in the binomial theorem and rearrange the resulting equation to deduce the following theorem:

> Let S be an n-element set, where $n > 0$. The number of subsets of S having an odd number of elements equals the number of subsets of S having an even number of elements.

14. Use a combinatorial argument to prove that

$$C(2n, n) = \sum_{k=0}^{n} [C(n, k)]^2.$$

3.4 Selections and Arrangements of Nondistinct Objects

TWO PROBLEMS

1. How many different eight-letter words can be formed from the letters in the word "arranger"?

2. Each player in a game of chess controls 16 pieces consisting of 8 pawns, 2 rooks, 2 knights, 2 bishops, 1 king, and 1 queen. Each piece except the king may be captured by the opposing pieces. How many recognizably different sets of pieces of one player can have been captured at any given moment during the game? ∎

Problem 1 asks for the number of *arrangements* of eight objects, some of which are identical. Problem 1 is a variation of the arrangement problem in Section 3.2, where all of the given objects were assumed to be different.

Problem 2, however, presents a variation of the selection problem in Section 3.3, which asks for the number of *selections* from a set of n elements that are classified into m types when we regard all elements of a given type as indistinguishable.

To solve Problem 1, we begin by listing the distribution of letters in the word "arranger"; the letter a occurs twice, the letter r occurs three times, and each of the other three letters occurs once. Initially it is convenient to distinguish between identical letters by assigning a different subscript to each such letter; thus the eight letters are written as $a_1, a_2, r_1, r_2, r_3, n, g$, and e. If we regard them as eight distinguishable letters, those letters can be arranged linearly in $P(8, 8) = 8!$ ways. One example of such an arrangement is

$$g \quad r_2 \quad a_1 \quad r_3 \quad n \quad a_2 \quad e \quad r_1. \tag{3.4-1}$$

Next, we abandon our pretense that repeated letters can be distinguished, which means that the subscripts can no longer be regarded as distinguishing symbols. In particular, arrangement (3.4-1) cannot be distinguished from any arrangement obtained by permuting identical letters, such as

$$g \quad r_2 \quad a_2 \quad r_3 \quad n \quad a_1 \quad e \quad r_1.$$

This example has one pair of identical letters that can be permuted in $P(2, 2)$ ways and one triple of identical letters that can be permuted in $P(3, 3)$ ways. Therefore, arrangement (3.4-1) is indistinguishable from the product of $P(2, 2) = 2!$ and $P(3, 3) = 3!$ arrangements of eight distinct letters. Thus, our initial count of $P(8, 8) = 8!$ counted each distinguishable arrangement $2!(3!)$ times instead of once. To compensate for that overcount, we compute the correct count by division:

$$\frac{P(8, 8)}{P(2, 2)P(3, 3)} = \frac{8!}{2! \, 3!} = 3360.$$

We now present this result formally.

Theorem 3.4-1 Let S be a set of n objects, each of which is one of m different types. Let $k_1, k_2, \ldots, k_m$ denote m positive integers such that k_i objects of S are of type i for $i = 1, 2, \ldots, m$ and $k_1 + k_2 + \cdots + k_m = n$. If all objects of each type are regarded as indistinguishable, then the number of arrangements of the n objects is

$$\frac{n!}{k_1! \, k_2! \, \cdots \, k_m!} .$$

Proof Consider a row of n empty boxes; into each box one object is to be placed to form an arrangement of the n objects. Two such arrangements will be recognized as different if and only if an object of some type appears in a position in one arrangement that is different from each position occupied by objects of that type in the other arrangement. We begin by choosing k_1 of the n empty boxes and assigning one object of type 1 to each such box. That can be done in $C(n, k_1)$ ways. Next we choose k_2 of the $n - k_1$ unfilled boxes and assign one object of type 2 to each such box. That can be done in $C(n - k_1, k_2)$ ways. By the Multiplicative Rule of And, the number of ways in which we can make the two sets of choices is

$$C(n, k_1)C(n - k_1, k_2) = \frac{n!}{k_1!(n - k_1)!} \cdot \frac{(n - k_1)!}{k_2!(n - k_1 - k_2)!}$$

$$= \frac{n!}{k_1!\, k_2!(n - k_1 - k_2)!}.$$

Continuing in this manner, we see that the m sets of choices can be made in

$$C(n, k_1)C(n - k_1, k_2)C(n - k_1 - k_2, k_3)$$
$$\cdots C(n - k_1 - k_2 - \cdots - k_{m-1}, k_m)$$

ways. Now if we replace each of these symbols by its factorial form, simplify the resulting expression by canceling like terms in the numerator and denominator, and use the relation

$$n - k_1 - k_2 - \cdots - k_{m-1} = k_m,$$

we obtain the number of arrangements stated in the theorem. ∎

Definition 3.4-1 If $\{k_1, k_2, \ldots, k_m\}$ is a set of positive integers whose sum is n, the *multinomial* symbol $C(n; k_1, k_2, \ldots, k_m)$ denotes the integer

$$\frac{n!}{k_1!\, k_2! \cdots k_m!}.$$

Observe that when we select k objects from n distinct objects, each selection of k objects also "selects" the remaining $n - k$ objects. Hence, the number $C(n, k)$ of selections of k objects from n objects can also be written in the form

$$C(n, k) = C(n; k, n - k) = \frac{n!}{k!(n - k)!},$$

which is the assertion of Theorem 3.4-1 for the binomial case, $m = 2$. For $m > 2$ the symbol

$$C(n; k_1, k_2, \ldots, k_m)$$

is called a *multinomial coefficient,* and its value is the coefficient of $x_1^{k_1} x_2^{k_2} \cdots x_m^{k_m}$ in the expansion of the expression $(x_1 + x_2 + \cdots + x_m)^n$.

EXAMPLE 1   In a newly elected state legislature there are 15 new members of the political party in power but only 12 committee appointments to be made: 2 on the finance committee, 3 on the budget committee, 4 on the education committee, and 3 on the agriculture committee. In how many ways can the appointments be made?

This is a selection problem of two stages. Initially 12 of the 15 new members must be selected for appointment to some committee. This selection can be done in $C(15, 12)$ ways. Then from those 12 members 2 must be assigned to finance, 3 to budget, 4 to education, and 3 to agriculture. This second stage can be done in $C(12; 2, 3, 4, 3)$ ways. Thus the number of ways of making the appointments is

$$C(15, 12)C(12; 2, 3, 4, 3) = \frac{15!}{12! \, 3!} \frac{12!}{2! \, 3! \, 3! \, 4!}$$

$$= \frac{15!}{2! \, (3!)^3 4!} = 126,126,000. \qquad \blacksquare$$

Selections of Subsets of Any Size

The question that we consider next is modeled by Problem 2, stated at the beginning of this section. Imagine that we have a set S of n objects that are of m distinguishable types: k_1 identical objects of type 1, k_2 identical objects of type 2, and so on, up to k_m identical objects of type m. How many recognizably different subsets does S have?

Theorem 3.4-2 Let S be a set of n elements, where each element is one of m distinguishable types such that all elements of a single type are identical. For $i = 1, 2, \ldots, m$ let the number of elements of type i be denoted by k_i, where $k_1 + k_2 + \cdots + k_m = n$. The number of distinguishable subsets of S of all sizes is

$$(k_1 + 1)(k_2 + 1) \cdots (k_m + 1).$$

Proof Any two subsets of S can be recognized as different if and only if for some index i the number of elements of type i in one set is different from the number of elements of type i in the other set. Let j denote an arbitrary but fixed index. Then each subset of S contains exactly 0, 1, 2, . . . , or k_j elements of type j. Thus, relative to type j elements, there are exactly $k_j + 1$ different kinds of subsets of S. Hence, relative to elements of all types, there are $(k_1 + 1)(k_2 + 1) \cdots (k_m + 1)$ recognizably different subsets of S. $\qquad \blacksquare$

EXAMPLE 2 Problem 2 asks for the number of recognizably different subsets of the 16 chess pieces that one player could legally have captured during the play of a game. Because the king cannot be captured, we can apply the previous theorem with $n = 15$, $k_1 = 8$, $k_2 = k_3 = k_4 = 2$, $k_5 = 1$, and $m = 5$. Hence, the number of distinguishable subsets (excluding all subsets that contain the king) is

$$(8 + 1)(2 + 1)(2 + 1)(2 + 1)2 = 2(3^5) = 486.$$

This number is only a small proportion of the number 2^{15} of all subsets of a set having 15 distinct elements. ■

Exercises 3.4

1. A school gym class consists of 30 students. The instructor wants to choose two volleyball teams of nine players each, two basketball teams of five players each, and two persons to referee (one for basketball and one for volleyball). Express in factorial form the number of ways in which the choices can be made.

2. The Central High basketball squad consists of 14 players, including 3 centers, 6 guards, and 5 forwards. In how many ways can the coach select a starting lineup that includes the team captain, who is a forward? (A team has 1 center, 2 guards, and 2 forwards.)

3. How many different arrangements can be made from the 11 letters of "abracadabra"?

4. Refer to Problem 1 at the beginning of this section and determine the number of different seven-letter words that can be formed from the eight letters in "arranger."

5. Perform all the algebraic details needed to complete the proof of Theorem 3.4-1.

6. A 5-person committee is to be formed from a campus senate of 15 persons (3 administrators, 5 faculty, and 7 students) consisting of 7 women and 8 men. How many different committees can be formed in each of the following circumstances?
   (a) The committee must contain at least 1 administrator and at least 3 students.
   (b) At least 2 of each sex must be included.
   (c) No administrator is to be included, but the committee cannot consist entirely of students.

7. In a certain midwest town the avenues run east and west and are numbered successively both north and south from Center Avenue. The streets run north and south and are numbered successively both east and west of Division Street.

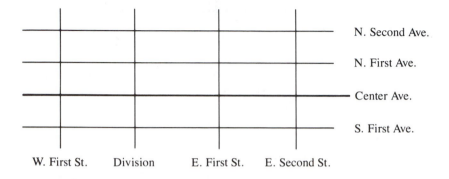

A pedestrian starts at Center and Division and walks to the corner of W. Tenth Street and S. Fourteenth Avenue, proceeding west or south at all times along the sidewalks that border the trafficways. How many different choices of routes does the pedestrian have? Explain your reasoning.

8. Expand the following expressions algebraically and observe how the coefficients occur from the expansion process:

   (a) $(x + y + z)^2$,
   (b) $(x + y + z)^3$.

   State a general trinomial theorem on the basis of those expansions. Your theorem should bear some resemblance to the binomial theorem and should be of the form

   $$(x + y + z)^n = \text{a sum of terms,}$$

   where each term in the sum is a product of the form $kx^ay^bz^c$, where k is a number depending on integers a, b, and c, whose sum is n.

9. (a) How many different 16-letter words can be created from the 16 letters MISSISSIPPI STATE? (Every arrangement of letters is a "word.")
   (b) How many 15-letter words can be formed if one of the vowels is omitted?

10. A person has just purchased four oranges, three grapefruit, and five apples and would like to take some or all of the fruit to a friend. How many recognizably different collections of fruit could be selected from that purchase?

11. How many integers $n > 1$ are divisors of 10,800?

12. Assume that a convex octagon is drawn so that no three of its interior diagonals intersect in a single point. Into how many line segments are the interior diagonals subdivided by their points of intersection?

3.5 Selections and Arrangements with Repetition

PROBLEM Ten lawyers are eating lunch together at a restaurant that offers seven different desserts. How many different selections of desserts can be ordered by the group? ∎

This problem asks for the number of recognizably different sets of n objects selected from arbitrarily large inventories of m different types of items, with all items of the same type regarded as identical. A corresponding question about arrangements is easier to answer, so we consider it before returning to the selection problem.

We can assign meaning to the concept of an arbitrarily large inventory either as the output of a production line that automatically produces another copy of an item as soon as that item is chosen (a toothpick dispenser in a restaurant) or as a process that chooses one object from a sample case of m objects (one of each type), records its type, and returns it to the sample case. The latter view explains why this variation of arrangement and selection is sometimes termed "with *replacement*," rather than "with repetition."

> **Theorem 3.5-1** Given a large supply of each of m distinct types of objects, the number of arrangements of n objects selected from those supplies is m^n.

> **Proof** Consider a line of n boxes, labeled from 1 to n. For each of the n boxes m types of objects can be assigned to that box; hence, there are m^n arrangements in all. ∎

EXAMPLE 1 How many 4-letter "words" are there in a language that uses one of the following alphabets: Arabic, German, Greek, Hebrew, Roman, Russian, Sanskrit (Devanagari).

In mathematical usage a "word" is any finite sequence of a given set of symbols. If repetition is permitted, m^n n-letter words can be formed with an m-letter alphabet. For this example, $n = 4$, and m depends on the alphabet used. For the alphabets named, m is, respectively, 28, 31, 24, 23, 26, 35, and 42, and the smallest and largest numbers of 4-letter words are

$$23^4 = 279{,}841 \text{ in Hebrew}$$
$$42^4 = 3{,}111{,}696 \text{ in Sanskrit.} \qquad \blacksquare$$

Selections with Repetition Permitted

This variation of the selection problem can be described as the "bakery shop" problem. We want to buy a dozen cookies, but when we see six tempting varieties available in the display case of the bakery we ask the clerk for two ginger cookies, four oatmeal cookies, three brownies, and one each of the three other varieties. Or perhaps we decide to buy a dozen brownies. How many orders are possible?

To formalize this question, we let m denote the number of types of objects, each available in virtually limitless supply. For $i = 1, 2, \ldots, m$ let k_i denote the number of objects of type i in a given selection, and let n denote the total number of objects selected: $n = k_1 + k_2 + \cdots + k_m$.

> **Theorem 3.5-2** The number of selections of n objects from a set of m distinct types of objects, if repeated selection of objects of the same type is allowed, is
>
> $$C(n + m - 1, m - 1).$$

Proof Think of the m types of objects as numbered from 1 to m. We want to count unrestricted selections of n objects, so we can think of n boxes to be filled with one object in each box. Without loss of generality we can agree to put all items of type 1 in the first k_1 boxes, all items of type 2 in the next k_2 boxes, from left to right, with each of the last k_m boxes containing an item of type m. And each k_i can be chosen arbitrarily provided that $k_1 + k_2 + \cdots + k_m = n$.

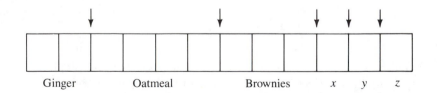

The sketch illustrates the particular choices made at the bakery shop. In general, however, we can think of $m - 1$ "dividers" that separate the row of n boxes into at most m segments. Each divider can be placed immediately before or after any box (including the possibility of two dividers at the extreme left and the remaining dividers at the extreme right, if we decide to buy a dozen brownies). Now think of each divider as a new box with the lid closed, to be placed within (or at either end of) the row of n boxes to make a row of $n + m - 1$ boxes, of which n are open and $m - 1$ are closed. Regarded in this way, we see that the number of selections of n objects from a set of m distinguishable types, allowing repeated selection of objects of the same type, equals the number of ways of choosing $m - 1$ boxes from a set of $n + m - 1$ boxes, $C(n + m - 1, m - 1)$. (Recall that the symbol $C(n + m - 1, n)$ describes the same number.) ■

EXAMPLE 2 Consider the problem at the beginning of this section. If each of the 10 lawyers orders dessert, there must be repetition, and the previous theorem can be applied with $n = 10$ and $m = 7$. Therefore, the number of distinguishable dessert orders for the group of lawyers is

$$C(10 + 7 - 1, 7 - 1) = 8008.$$

If we want to allow the possibility that some persons might order no dessert, then there are 8 types of individual dessert orders, and we have $m = 8$ and $n = 10$. Hence,

$$C(10 + 8 - 1, 8 - 1) = 19,448$$

is the number of different dessert orders for the group. ■

EXAMPLE 3 How many different nine-letter words can be formed from the letters MISSISSIPPI if M and P must each appear at least once and I and S must each appear at least twice?

We know that we must use at least one M, one P, two I's, and two S's. That leaves one P, two I's, and two S's from which we need to choose three additional letters. We could choose

(a) one P, one I, and one S, or

(b) two I's and one P, or

(c) two S's and one P, or

(d) two I's and one S, or

(e) two S's and one I.

Observe that **b** and **c** are alike except for an interchange of I and S, and in the same manner **d** and **e** are alike. Hence, we will count the number of words formed for cases **a**, **b**, and **d**. The number of words for cases **c** and **e** will be the same, respectively, as for **b** and **d**. For each case we need only count the number of different arrangements of the nine chosen letters.

(a) $C(9; 3, 3, 2) = \dfrac{9!}{(3!\ 3!\ 2!)} = 5040,$

(b) $C(9; 4, 2, 2) = \dfrac{9!}{(4!\ 2!\ 2!)} = 3780,$

(d) $C(9; 4, 3) = \dfrac{9!}{(4!\ 3!)} = 2520.$

Hence, the total number is $5040 + 2(3780 + 2520) = 17{,}640.$ ∎

EXAMPLE 4   A person is taking a basket of 12 pieces of fruit to a sick friend. The basket is to contain bananas, oranges, grapefruit, and apples with at least 1 piece of each type of fruit. How many different baskets of fruit can be prepared?

This problem is one of selection with repetition permitted, with the restriction that each of 4 fruits must be included. Think of putting 1 piece of each type of fruit in the basket and then deciding how to select the remaining 8 pieces. Then there will be 8 open boxes to fill and 3 closed boxes to introduce as dividers. Hence, the number of different ways to complete the task of filling the basket is $C(8 + 4 - 1, 3) = 165.$ ∎

Exercises 3.5

1. The news vendor in an office building stocks Hershey bars, Heath bars, and Milky Ways. Ten of these items were sold during lunch hour on Tuesday.

   (a) How many different arrangements of ten such candy bars are possible?
   (b) How many arrangements are possible if no two consecutive candy bars are of the same brand?

2. A florist with a large supply of yellow daffodils, blue irises, and white narcissus is asked to deliver a dozen of those flowers to a woman on her ninety-first birthday. How many different bouquets are possible?

3. How many four-digit sequences can be formed from the digits 1 through 9 given that

   (a) the sequence is strictly increasing,
   (b) the sequence is nondecreasing,
   (c) no restriction is imposed on the sequence?

4. A panel of 11 persons (students, faculty, and administrators) is to study a proposed change in the academic calendar. In how many ways can the panel be constituted (relative to the number of representatives of each of those three groups) if

   (a) at least 2 members of each of the 3 groups must serve on the panel,
   (b) at least 3 members of each group are on the panel,
   (c) the chair must be an administrator, and at least 4 students must be on the panel?

5. How many six-digit natural numbers are divisible by 5?

6. To earn his allowance, a boy is expected to perform duties at home. Each month he is to complete 20 half-hour jobs that he chooses (repetition allowed) from washing dishes, walking the dog, feeding the chickens, weeding the garden, or painting the basement walls. In how many ways can he select 20 jobs?

7. If you are allowed to choose any seven coins from a roll of pennies, a roll of nickels, and a roll of dimes, how many different collections could you choose? (Regard all coins of the same denomination to be identical.)

8. In Exercise 7 suppose you are allowed to choose seven of the coins but no more than one nickel nor more than four pennies. How many different sums of money can be obtained by choosing seven coins?

9. Given the seven letters ALI BABA, how many words can be formed having

   (a) seven letters,
   (b) six letters,
   (c) five letters?

10. From Exercise 9 we observe that the number of six-letter words that can be formed from the seven given letters equals the number of seven-letter words that can be formed from those letters. Show that equality holds generally for the numbers of k-letter and $(k-1)$-letter words formed from any given set of k letters.

11. Eight photographers went on a 12-day safari into tiger country and slept outdoors for the 11 nights. During the first 8 nights, each person kept watch through 1 night. For each of the last 3 nights the watch was kept by the person who had drawn the shortest straw that afternoon, excluding the person who had kept watch on the previous night. How many watch schedules are possible for the 11-night period?

12. Given the 11 letters of "abracadabra,"

    (a) how many different 6-letter selections can be made?

    (b) how many different 6-letter words can be formed?

3.6   Distributions: Assignments

THREE PROBLEMS WITH ONE SOLUTION

1. In how many ways can 12 pieces of fruit be selected from large supplies of apples, bananas, and cherries?

2. In how many ways can 12 lemons be assigned to three different baskets?

3. How many ordered triples (a, b, c) of natural numbers satisfy the equation $a + b + c = 12$?   ■

We recognize the first problem as an example of selection with repetition, counting the number of selections of 12 objects from supplies of three different types. By Theorem 3.5-2 the answer is

$$C(12 + 3 - 1, 2) = C(14, 2) = 91.$$

The second problem asks for the number of ways in which 12 identical objects can be sorted into three different baskets. Imagine having a large supply of labels of three types: for example, "apples," "bananas," and "cherries." First, place one of each type of label on some basket to distinguish the baskets from one another; then place a label on each of the 12 lemons, choosing the labels in any way; finally, sort the labeled lemons into the corresponding baskets. In this manner 12 pieces of fruit have been placed in three baskets labeled apples, bananas, and cherries; this corresponds to one way in which the selection described in Problem 1 can be carried out. Conversely, any way of selecting fruit as specified in Problem 1 defines a way to distribute 12 lemons among three baskets. Hence, Problem 2 also must have 91 as its answer.

Problem 3 is simply an algebraic restatement of Problem 1, because we can interpret the letter a to be the number of apples, b the number of bananas, and c the number of cherries. Hence, there are 91 different ordered sets of three natural numbers that add to 12. The word "ordered" is necessary in the previous sentence because three apples, two bananas, and seven cherries is a different selection from seven apples, two bananas, and three cherries.

Theorems 3.6-1 and 3.6-2 answer the questions posed in Problems 3 and 2, respectively.

Theorem 3.6-1 Let m and n be natural numbers. The number of ordered m-tuples $(x_1, x_2, \ldots, x_m)$ of natural numbers such that $x_1 + x_2 + \cdots + x_m = n$ is

$$C(n + m - 1, n).$$

Proof See Exercise 11. ∎

We can interpret Theorem 3.6-1 as an observation that the number of ordered m-tuples of natural numbers whose sum is n is $C(n + m - 1, n)$. For $n = 4$ and $m = 3$ we have $C(6, 4) = 15$; the 15 sequences are

4, 0, 0	3, 1, 0	2, 2, 0
0, 4, 0	3, 0, 1	2, 0, 2
0, 0, 4	1, 3, 0	0, 2, 2
	1, 0, 3	2, 1, 1
	0, 3, 1	1, 2, 1
	0, 1, 3	1, 1, 2

Thus Theorem 3.6-1 provides a formula for counting all *ordered* representations of n as a sum of m *nonnegative* integers. That counting problem must not be confused with the following problem: Determine the number of *unordered* sets of *positive* integers whose sum is n. An unordered set of positive integers whose sum is n is called a *partition* of n. A partition of n can also be regarded as a nonincreasing sequence of positive integers whose sum is n. For example, there are five partitions of 4:

$$(4) \quad (3, 1) \quad (2, 2) \quad (2, 1, 1) \quad (1, 1, 1, 1).$$

Methods for counting partitions of n are presented in many of the references on combinatorics.

Problem 2 is a specific example of various combinatorial problems called *distributions*. It asks us to find the number of ways of assigning n identical objects (12 lemons) to m different boxes (3 baskets). Other forms of distributions regard the objects as distinct (rather than identical) or the boxes as identical (rather than distinct). Still other types of distributions require arrangement of the objects (rather than assignment), or perhaps they specify that none of the boxes may be empty. For clarity and convenience we summarize all these types of distributions by formulating a single statement that incorporates two alternatives in each of four positions within the statement.

GENERAL
DISTRIBUTION
PROBLEM

Determine the number of ways in which

$n \begin{cases} \text{identical} \\ \text{distinct} \end{cases}$ objects can be $\begin{cases} \text{assigned to} \\ \text{arranged within} \end{cases}$

$m \begin{cases} \text{identical} \\ \text{distinct} \end{cases}$ boxes, with empty boxes $\begin{cases} \text{allowed} \\ \text{not allowed} \end{cases}$

Initially it appears that 2^4 different problems can be stated by choosing each of the two alternatives for the four positions in braces in this statement. However, all arrangements of identical objects are indistinguishable, so four of the eight arrangement problems coincide with the four corresponding assignment problems, which reduces the number of different problems to 12. This section shows how to solve three of the assignment problems; Section 3.7 is devoted to two of the four arrangement problems. The remaining distribution problems can be solved by more advanced techniques, some of which are introduced in Sections 3.9 and 3.10; several examples of such problems are discussed briefly in Section 3.7.

> **Theorem 3.6-2**   The number of ways in which n *identical* objects can be *assigned* to m *distinct* boxes with empty boxes *allowed* is
>
> $$C(n + m - 1, n).$$

Proof   For $i = 1$ to m let x_i denote the number of objects to be assigned to box i. Then $x_1 + x_2 + \cdots + x_m = n$.

Conversely, each solution of that equation specifies the number of objects to be assigned to box i for each i. Also, two solutions are equal if and only if the corresponding assignments are the same, so Theorem 3.6-2 follows from Theorem 3.6-1. ■

From Theorem 3.5-2 we recall that $C(n + m - 1, n)$ also specifies the number of ways in which n objects can be selected with repetition from unlimited supplies of each of m different types of objects. Thus, assigning k identical objects to the same box in Theorem 3.6-2 can be regarded as selecting k objects of a given type in Theorem 3.5-2.

EXAMPLE 1   Thirty-six purple hard-boiled eggs are scattered across a large lawn for children to find in the annual egg hunt. If nine children set out on the hunt and find all the eggs, in how many ways can the eggs be distributed in the baskets when the children return?

We regard the eggs as identical but the children as distinct. Hence, Theorem 3.6-2 can be applied to obtain the answer

$$C(36 + 9 - 1, 36) = C(44, 36) = 177,232,627. \qquad \blacksquare$$

EXAMPLE 2   Suppose the previous example specified that the child with the most eggs must give 1 egg to a child who finds fewer than 2 eggs, with this process being continued (the child with the most giving 1 egg to the child with the least until each child has at least 2 eggs). In how many ways can the eggs be distributed?

To answer this question, we simply pretend that each child starts with 2 eggs out of the original 36. Hence, the 9 children will have only $36 - 2(9)$ eggs to hunt for. In how many ways can 18 identical eggs be assigned to 9 distinct baskets? From Theorem 3.6-2 the answer is

$$C(18 + 9 - 1, 18) = C(26, 18) = 1,562,275. \qquad \blacksquare$$

Theorem 3.6-3   The number of ways in which n *identical* objects can be *assigned* to m *distinct* boxes with empty boxes *not allowed* is

$$C(n - 1, m - 1).$$

Proof   We first use the method illustrated in the preceding example; to ensure that no box is empty, we first place 1 object in each of the m boxes. Then we have $n - m$ objects to be assigned to m distinct boxes in all possible ways. According to Theorem 3.6-2, the number of ways of doing that is

$$\begin{aligned} C((n - m) + (m - 1), n - m) &= C(n - 1, n - m) \\ &= C(n - 1, m - 1), \end{aligned}$$

where the last equality follows by the symmetry property of $C(p, k)$, namely,

$$C(p, k) = C(p, p - k). \qquad \blacksquare$$

Theorem 3.6-4   The number of ways in which n *distinct* objects can be *assigned* to m *distinct* boxes with empty boxes *allowed* is m^n.

Proof   See Exercise 13. $\qquad \blacksquare$

Observe that the boxes in Theorem 3.6-4 are assumed to be distinguishable, as are the articles to be assigned to the boxes, so a new assignment is created if the contents of two boxes are interchanged, for example. Thus, an assignment of one object to each of m distinguishable boxes is an arrangement of those objects, and different assignments will be distinguishable whenever the m objects are distinguishable. However, Theorem 3.6-4 allows some boxes to be empty and some boxes to contain several objects. The former must occur if $m > n$, the latter must occur if $n > m$, and both are permitted to occur regardless of the values of m and n. The m boxes can be regarded as being labeled from 1 to m. Suppose that each object in box i is also regarded as being labeled i. Then all articles in box i bear indistinguishable labels, and we can regard box i as providing a set of identical labels with a total of n labels in the m boxes. Therefore, any assignment of the form described in Theorem 3.6-4 determines a means of choosing a total of n copies of m types of labels and assigning one chosen label to each of n distinct objects. Thus, we can calculate the desired number of assignments as we did in Theorem 3.5-1.

EXAMPLE 3   The number of 10-digit numerals that can be formed in base-8 notation (see Exercises 17 through 20 of Section 1.9) is 8^{10}, because we can regard each position of a 10-digit numeral as a box into which any 1 of 8 digits can be placed. Observe that $8^{10} = (2^3)^{10} = (2^{10})^3$; because $2^{10} = 1024$, we see that 8^{10} is larger than 1 billion.   ∎

Exercises 3.6

1. A cafeteria offers five types of sandwiches at a price of $2 each. How many different selections of sandwiches can a customer obtain for $20?

2. In how may ways can a storekeeper make change for a customer who has just paid for a $4.63 purchase with a $5 bill?

3. Determine the number of different solutions in positive integers of the inequality $x_1 + x_2 + x_3 < 50$.

4. Newsstand attendants in airports often are asked to change a $1 bill into some combination of quarters, dimes, and nickels. In how many ways can it be done?

5. In how many ways can a four-letter word be formed from a large supply of each of the six letters in "mother"?

6. How many positive integers less than 10 million have no even digits?

7. A jar contains red, white, black, yellow, and green jelly beans, all in ample supply. How many different color combinations can be obtained by choosing 10 jelly beans?

8. At a child's birthday party 16 identical favors are to be given to the 16 invited guests, but only 11 guests attend the party.
   (a) In how many ways can 16 favors be distributed to 11 children?
   (b) In how many ways can the 16 favors be distributed to the 11 guests so that each guest gets at least one favor?

9. In planning next year's budget, a family set aside $1000 for donations to five types of organizations: religious, educational, health care, community support, and public welfare. If at least $100 is to be allocated to each type of organization, in how many ways can a total of $1000 be allocated?

10. How many positive integers divide evenly into 122,850?

11. Prove Theorem 3.6-1.

12. An instructor offers his class a choice of three possible test dates, with the actual date to be chosen by a majority vote of the 31 class members.
   (a) In how many ways can the 31 votes be distributed among the three dates?
   (b) In how many ways can the votes be distributed to produce a decision?

13. Prove Theorem 3.6-4.

3.7 Distributions: Arrangements

We now turn our attention to the two forms of the General Distribution Problem that ask for the number of different ways in which n distinct objects can be arranged within m distinct boxes; empty boxes may or may not be allowed.

Theorem 3.7-1 The number of ways in which n *distinct* objects can be *arranged* within m *distinct* boxes with empty boxes *allowed* is

$$P(n, n)C(n + m - 1, m - 1).$$

Proof Think of arranging the n objects in a line, which can be done in $n! = P(n, n)$ ways. To place these n objects in m or fewer boxes without disturbing the order in which they are arranged, we can imagine introducing $m - 1$ dividers (closed boxes) anywhere along the line. Regard all the objects to the left of the first divider as being in box 1, all of the objects between the first and second dividers as being in box 2, and so on. As in Theorem 3.5-2, the number of ways in which this can be done is $C(n + m - 1, m - 1)$. Thus the number of ways of arranging the n objects and placing m box-defining dividers is

$$P(n, n)C(n + m - 1, m - 1) = n! \frac{(n + m - 1)!}{n!(m - 1)!}$$
$$= \frac{(n + m - 1)!}{(m - 1)!}.$$

Because of the symmetry of the binomial coefficients, we can write this number as $n!C(n + m - 1, n)$.   ∎

EXAMPLE 1   In a naval convoy visual messages can be sent by signal flags by hoisting 1 or more of a set of distinctly colored flags on 1 or more masts of a ship. Different messages are created by changing either the selection or the order of the flags on any mast. If 10 distinct flags are available to be hoisted onto 3 masts, and if each mast can hold 10 or fewer flags, the number of different 10-flag messages that can be sent can be determined by Theorem 3.7-1 to be

$$10! \, C(10 + 3 - 1, 3 - 1) = 10! \, C(12, 2) = 239,500,800.$$

The flags can be arranged linearly in 10! ways, and then the placement of 2 dividers will indicate which set of arranged flags goes on each of the 3 masts. For example, if both dividers are placed before the first of the 10 flags, the flags all are to go on the third mast. If the dividers appear after the fourth and ninth flags, the first four flags appear on the first mast, the last flag goes on the third mast, and the middle set goes on the second mast.   ∎

Theorem 3.7-2 The number of ways in which n *distinct* objects can be *arranged* within m *distinct* boxes with empty boxes *not allowed* is

$$P(n, n)C(n - 1, m - 1).$$

Proof See Exercise 2. Recall the proofs of Theorems 3.6-3 and 3.7-1.   ∎

EXAMPLE 2 In the previous example, if all 3 masts must be used for a 10-flag signal, Theorem 3.7-2 applies and yields the number

$$10! \, C(9, 2) = 130{,}636{,}800.$$

It follows that the number of 10-flag signals that can appear on 2 or fewer masts is

$$10! \, [C(12, 2) - C(9, 2)] = 10!(30) > 10^8. \qquad \blacksquare$$

Examples 3 through 6 discuss two specific assignment problems and two specific arrangement problems not covered by the theorems of Sections 3.6 and 3.7.

EXAMPLE 3 In how many ways can five identical objects be assigned to four identical boxes, empty boxes being allowed?

Because the objects are identical, the problem is to find the number of ways in which four natural numbers can be selected to have a sum equal to 5. For example, $2 + 0 + 3 + 0 = 5$; this sum corresponds to an assignment in which one box is assigned two objects, another box is assigned three objects, and the remaining two boxes are left empty. Because the boxes are identical, the order in which these four numbers are given does not matter; to avoid writing any set of four natural numbers in two different orders, we shall write the numbers as a nonincreasing sequence, $(3, 2, 0, 0)$ in this instance. The question in this example now becomes: In how many ways can a nonincreasing sequence of four natural numbers be formed such that its sum is 5? We can form exactly six such sequences.

$(5, 0, 0, 0)$	$(4, 1, 0, 0)$	$(3, 2, 0, 0)$
$(3, 1, 1, 0)$	$(2, 2, 1, 0)$	$(2, 1, 1, 1)$

Therefore we conclude that there are six ways in which five identical objects can be assigned to four identical boxes. $\blacksquare$

As noted in Section 3.6, each of the six sequences listed is called a *partition* of the natural number 5 into four or fewer parts. More generally, a partition of a positive integer n into m or fewer parts is a nonincreasing sequence of m natural numbers whose sum is n. The problem of determining the number of such partitions for each pair of positive integers m and n with $m < n$ is discussed in many of the books on combinatorics listed in the References.

EXAMPLE 4   If we modify Example 3 by stipulating that empty boxes are not allowed, we can rephrase the question as follows: How many nonincreasing sequences of four positive integers have a sum equal to 5? From Example 3 the only such sequence is (2, 1, 1, 1), so the answer is 1.   ∎

EXAMPLE 5   In how many ways can five distinct objects be arranged within four identical boxes if empty boxes are allowed?

Let the five distinct objects be distinguished only by labels numbered from 1 to 5. Imagine removing the labels so that the objects are now identical. Assign those five identical objects in any way among the four identical boxes; by Example 3 that can be done in six ways. Without disturbing the assignment of objects within boxes, we see there are $P(5, 5) = 5!$ ways of pasting the labels from 1 to 5 on those five objects. Thus there are $6(5!) = 720$ ways of arranging five distinct objects within four identical boxes.   ∎

EXAMPLE 6   If we now modify Example 5 by stipulating that empty boxes are not allowed, we can use Examples 4 and 5 to conclude that there are now $1(5!) = 120$ ways to arrange five distinct objects within four identical boxes if empty boxes are not allowed.   ∎

Exercises 3.7

1. Suppose you write the nine nonzero digits in any order and then insert a hyphen between each of any three pairs of adjacent digits to form an ordered set of four numerals: for example, 37-2156-8-94.

   (a) How many ordered sets of four numerals can be formed in this way from the nine nonzero digits?

   (b) How many ordered sets of fewer than four numerals can be formed in this way by using fewer than three hyphens as separators?

2. Prove Theorem 3.7-2.

3. A hunger committee collected 32 different cans of food to distribute to some or all of five distinct needy families. In how many ways can that food be parceled out to those families, given that no restrictions are imposed on the form of distribution?

4. Twelve students are to be scheduled for individual half-hour conferences with the dean during a period of five days. For each of the following cases, determine the number of different orders in which

the conferences can be arranged within the five days. (Two orders are considered different if either the students are ordered differently or if the numbers of conferences scheduled on a given day are different.)

(a) No restrictions are imposed.

(b) At least one conference must be scheduled each day.

(c) At least two but no more than three conferences must be held daily.

5. A numismatist keeps a valuable collection of six rare and distinct Greek coins in three identical storage cases. Each case can store six coins in a line along the length of the case. Use Example 5 as a guide to answer the following questions.

(a) In how many different ways can the collection be stored in the cases if empty cases are allowed?

(b) In how many different ways can the collection be stored in two or fewer cases?

6. (a) In how many different ways can 6 identical lemons be placed in 4 identical baskets?

(b) In how many ways can 14 identical lemons be placed in 4 identical baskets if each basket is to contain at least 2 lemons?

7. Seven different kinds of candy are to be arranged in four identical boxes. Each box can hold up to seven pieces of candy, arranged in a line along the length of the box, and the boxes are then placed along that line. In how many ways can the candy be arranged within the aligned boxes, given that

(a) no restrictions are imposed on the form of distribution;

(b) each box must contain at least one piece of candy?

8. In Example 1 suppose that 1 mast is shorter than the other 2 and can hold at most 5 flags.

(a) How many different 10-flag messages can be sent?

(b) How many different 10-flag messages can be sent if there must be at least 2 flags on each mast?

9. An instructor is to make up an examination in which the questions are to be separated into part A, part B, and part C. In how many ways can 24 questions be arranged and assigned to the 3 parts, given that

(a) each part must contain at least 5 questions;

(b) no part may contain more than 13 questions;

(c) each part must contain exactly 8 questions?

10. A string quartet has prepared 18 compositions for its concert tour, which includes 3 engagements over a single weekend in the Chicago area. The quartet agrees that no work should be played more than

once in a given area, and that each concert will be an ordered list of 4 compositions. How many different concerts can be arranged for the 3 Chicago appearances?

11. Explain why (or why not) Theorem 3.7-1 should be called "The Santa Claus Theorem."

FURTHER COUNTING TECHNIQUES

3.8 Recursive Counting

PROBLEM   In how many ways can one spell "NORTH CAROLINA" by proceeding steadily downward through the array in Figure 3.8-1?

Figure 3.8-1   ■

There are numerous ways to approach this question; an inexperienced problem solver might choose to follow the advice the King gave to the White Rabbit:

Begin at the beginning and go on until you come to the end: then stop.

But sometimes it is better to reverse this bit of wisdom: We can go to the end and look back to discover how we might have arrived there. And sometimes it is best to go to some intermediate position and ask how we could have reached that point. For example, in Figure 3.8-1, to reach the letter C we must first reach the letter H. If a C position is not on an edge of the array, that C can be reached only by first reaching either of the two H positions that flank C in the row above. If the C position is on the edge of the array, then only one preceding H position is possible. Hence, if we know the number of ways in which we can reach each H position, we can determine the number of ways in which we can reach any given C position.

The counting strategy described by the previous sentence illustrates the concept of *recursive counting.* Suppose we want to count the number of different ways in which a given process can be carried out; suppose also that the process can be regarded as the successive completion of a number of tasks, and that the number $t(k)$ of ways of completing task k can be calculated from the numbers $t(k-1), t(k-2), \ldots, t(k-r)$ for some fixed $r < k$. Then we can write an equation of the form

$$t(k) = f(t(k-1), t(k-2), \ldots, t(k-r), k),$$

where f is a function of $r + 1$ variables. An equation of this form is called a *recursion equation.* We have already seen numerous instances of recursive reasoning, as restated in Examples 1 through 4.

EXAMPLE 1   The factorial function $n!$ is defined for each natural number n by the starting value $0! = 1$ and the recursion rule $k! = k(k-1)!$.   ∎

EXAMPLE 2   The Tower of Hanoi puzzle with n disks (see Section 3.1 and Example 3 of Section 1.10) can be solved by counting moves in three stages: First, move the top tower of $n-1$ disks from peg 1 to peg 2, using $t(n-1)$ individual moves; next, move the largest disk to peg 3; finally, transfer the tower of $n-1$ disks from peg 2 to peg 3. Hence, a recursion equation for the puzzle is

$$t(n) = t(n-1) + 1 + t(n-1),$$

and the starting information is $t(1) = 1$.   ∎

EXAMPLE 3   From Example 4 of Section 3.1 we recall the plight of the tall basketball player who could climb steps only by taking 2 or 3 at a time. Thus to reach step k, the player must first reach either step $k-2$ or step $k-3$. Then

$$t(k) = t(k-2) + t(k-3),$$

with $t(0) = t(2) = 1$, $t(1) = 0$ as the starting data.   ∎

EXAMPLE 4   The Fibonacci sequence is described in Exercise 12 of Section 1.10 by an initial segment of the sequence together with the recursion equation

$$F(k + 1) = F(k) + F(k - 1), \quad k > 1.$$

The two initial values needed to start the sequence are $F(1) = F(2) = 1$. ∎

EXAMPLE 5   If n lines are drawn on the plane in such a way that no two lines are parallel and no three lines intersect in a single point, determine the number of regions into which the lines divide the plane.

The first line separates the plane into two regions, and the first two lines separate the plane into four regions. When the third line is drawn, the point of intersection of line 1 and line 2 must lie on one side of line 3, which therefore subdivides only three of the four previous regions of the plane. Also two new points of intersection are created, making three points of intersection and seven regions. Line 4 must intersect each of the previous lines, adding three new points of intersection (six in all) and adding four new regions (one before line 4 intersects any line, and one just after line 4 intersects each of the previous lines). These results are recorded in Table 3.8-1. A recursion equation for the number $t(n)$ of regions formed by n lines is therefore

$$t(n) = t(n - 1) + n. \quad ∎$$

From our initial discussion of the North Carolina problem, we note the presence of a recursion equation, and we sense that this question is somehow related to Pascal's triangle (Section 3.3) and thus to the binomial coefficients. To see the connection, we apply a technique, introduced by George Pólya, of assigning special coordinates to each position in the array. In Figure 3.8-1 think of the diagonal files of letters leading from the upper left margin of the array to the lower right margin as denoting streets numbered from 0 through 6, as in Figure 3.8-2. Also think of the diagonal files of letters leading from the upper right margin of the array to the lower

n	Points of Intersection	Regions of the Plane
0	0	$0 + 1 = 1$
1	0	$1 + 1 = 2$
2	1	$2 + 2 = 4$
3	$1 + 2 = 3$	$4 + 3 = 7$
4	$3 + 3 = 6$	$7 + 4 = 11$

Table 3.8-1

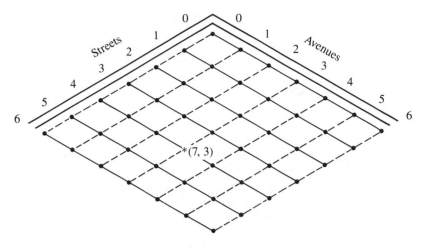

Figure 3.8-2

left margin as representing avenues. The Pólya coordinates that we assign to the intersection of street j and avenue k are $(j + k, k)$ for $0 \le j \le 6$ and $0 \le k \le 6$. If we walk from the top vertex, having Pólya coordinates $(0, 0)$, to the point in the array that has Pólya coordinates (r, s), then the total number of blocks that we have walked is r, which is also the number of intersections we have reached since leaving $(0, 0)$. At each such intersection we had to decide whether to walk southeast along a street or southwest along an avenue. The second coordinate s is the number of blocks we walked along streets, and $r - s$ is the number of blocks we walked along avenues.

Then for each pair (r, s) of natural numbers, such that both $0 \le s \le 6$ and $0 \le r - s \le 6$, we denote by $t(r, s)$ the number of different steadily descending routes in the array from the top vertex, with Pólya coordinates $(0, 0)$, to the position having Pólya coordinates (r, s). In terms of our earlier analysis, we can now write the following recursion equation for the number of different routes:

$$t(n, k) = t(n - 1, k - 1) + t(n - 1, k)$$

for all k and n such that $0 < k \le 6$ and $0 < n - k \le 6$, with initial conditions $t(m, 0) = 1 = t(m, m)$ for all m such that $0 \le m \le 6$. But this equation is the same as the recursion equation satisfied by the binomial coefficients $C(n, k)$. Furthermore, the two sets of initial conditions are the same:

$$C(m, 0) = 1 = t(m, 0) \quad \text{and} \quad C(m, m) = 1 = t(m, m) \quad \text{for } 0 \le m \le 6.$$

The shared recursion equation then guarantees that

$$t(n, k) = C(n, k) \qquad \text{for all allowed values of } k \text{ and } n.$$

Thus the number of descending paths that spell "North Carolina" is $C(12, 6) = 924$.

EXAMPLE 6 The streets of a South Dakota town run east and west. Main Street marks the east–west center line, and the streets numbered 1, 2, . . . run to the north and also to the south from Main. Similarly, avenues run north and south and are numbered in succession east and west from Center Avenue. A person gets off the bus at Center and Main and walks to East Tenth Street at South Twelfth Avenue. How many different routes of shortest distance are there?

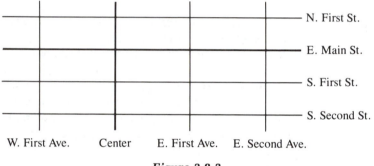

Figure 3.8-3

This problem is easily recognized as a barely disguised form of the Pólya spelling problem. Simply rotate Figure 3.8-3 clockwise through 45° and ignore all of the town that is either north of Main or west of Center to see that the two problems are identical. Hence, the number of routes will be $C(10 + 12, 12) = 646{,}646$. Alternatively, the problem can be solved by observing that a shortest path from Main and Center to East Tenth and South Twelfth is $10 + 12$ blocks long. The person begins by choosing whether to walk east or south one block, and the same decision is repeated at each corner until either 10 decisions to walk south or 12 decisions to walk east have been made. Whatever shortest route is taken, there are 22 corners, and the walker must choose 12 of them as points at which to walk east, so there are $C(22, 12)$ routes in all. ∎

As a final example, we consider an extension to three-dimensional space of Example 6.

EXAMPLE 7 Figure 3.8-4 represents a regular rectangular grid in three-dimensional space constructed of sections of pipe with threaded connectors at each junction. Coordinates are assigned to each junction point, with $(0, 0, 0)$ being the coordinates of the rear bottom corner O and (d, e, f) being the coordinates of the point P that is reached by moving from the origin d units along the positive x-axis, then e units parallel to the positive y-axis, and then f units directly upward. Along how many different shortest routes can an ant walking along the grid framework travel from O to P?

   The shortest route will be $d + e + f$ units long. At each junction point the ant must decide in which direction to travel to the next junction point; it can choose to travel in the positive direction of the x-axis, the y-axis, or the z-axis. For a shortest route the ant must choose the x direction d times, the y direction e times, and the z direction f times. Let $n = d + e + f$ be the number of decision points along any shortest route. To move d units in the x direction, the ant can choose any d of the n decision points along any shortest route in $C(n, d)$ ways. Then the ant can choose any e of the remaining $n - d$ decision points for a sequence of 1-unit moves in the y direction in $C(n - d, e)$ ways. Now there are only $n - (d + e) = f$ deci-

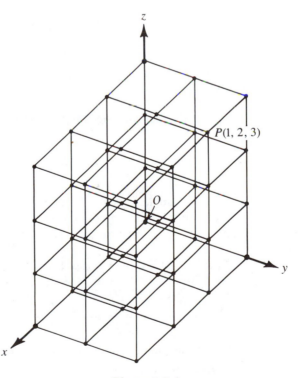

Figure 3.8-4

sion points unselected, and at each such point the ant must move 1 unit upward. Hence, the number of possible routes is

$$C(n, d)C(n - d, e)C(n - d - e, f).$$

The last of these three numbers is 1, of course.   ■

As an exercise, we may show that when the trinomial expression $(x + y + z)^n$ is expanded, then each term will be of the form

$$T(n; d, e, f)x^d y^e z^f,$$

where $d + e + f = n$. The coefficient $T(n; d, e, f)$ is called a *trinomial coefficient,* and it can be used to state a trinomial theorem that is analogous to the binomial theorem. The value of $T(n; d, e, f)$ is

$$C(n, d)C(n - d, e)C(n - d - e, f),$$

which we may show to be equal to

$$T(n; d, e, f) = \frac{n!}{d!\, e!\, f!}.$$

Exercises 3.8

1. Verify algebraically the claim of the last sentence of this section:

$$T(n; d, e, f) = \frac{n!}{d!\, e!\, f!}.$$

2. Use the jungle gym model of Example 7 to explain the following recurrence relation model for defining the trinomial coefficients: If d, e, and f, are positive integers such that $d + e + f = n$, then

$$T(n; d, e, f) = T(n - 1; d - 1, e, f)$$
$$+ T(n - 1; d, e - 1, f) + T(n - 1; d, e, f - 1).$$

3. Determine all initial conditions needed to accompany the recurrence equation of Exercise 2.

4. Use the algebraic representation of $T(n; d, e, f)$ to establish the result of Exercise 2.

5. Write a recursion equation for the number $s(n)$ of distinguishable sequences of n binary digits such that no two ones occur in adjacent positions. Determine the values of $s(n)$ for $1 \leq n \leq 8$.

6. A woman invests \$10,000 on July 1 at 10% interest, which is added to the account each year on July 1. At the end of the first year she withdraws \$100 on July 1, and each year thereafter she increases her withdrawal by \$100 on that date.

   (a) Write a recursion equation for the amount $A(n)$ that remains in her account on July 1, n years after the original investment. Assume that interest for the past year has been added and that her annual withdrawal has been deducted.

   (b) Calculate $A(n)$, rounding to the nearest dollar, for $1 \leq n \leq 5$.

7. Assume that a set of n ovals can be drawn in the plane so that each pair of ovals intersects in two distinct points, but no three ovals pass through the same point.

   (a) Write a recursion equation for the number $R(n)$ of nonoverlapping regions of the plane that are defined by the boundaries of those ovals.

   (b) Calculate the value of $R(k)$ for $k = 0, 1, 2, \ldots, 6$.

8. A narrow strip of fudge $n + 1$ inches long is marked at 1-inch intervals. The strip is to be sliced into $n + 1$ pieces in successive stages as follows: The first stage consists of a single slice that produces a left-hand piece of length k inches, for some $k = 1, 2, \ldots, n$, and a right-hand piece of length $n + 1 - k$. In each succeeding stage other slices are made from left to right along the strip, with precisely 1 slice being made in each uncut piece whose length at the end of the previous stage exceeded 1 inch. The process ends whenever the nth slice has been made. Let $S(n)$ denote the number of different ways in which n slices can be made in the given manner.

   (a) By considering all possible ways in which a single cut can be made in a strip of length $k + 1$, write a recursion equation for $S(n)$.

   (b) Noting that $S(0) = 1 = S(1)$, calculate $S(k)$ for $k = 2, 3, 4, 5$.

9. For a circle with $2n$ distinct points chosen around its circumference, let $P(n)$ denote the number of ways in which n pairs of the given points can be chosen so that no two chords joining the n pairs of points intersect in or on the circle. Assume that $n \geq 1$.

   (a) Write a recursion equation for $P(n)$.

   (b) Calculate $P(5)$.

   (c) Compare your results with those of Exercise 8.

3.9   The Concept of a Generating Function

We begin by recalling that the binomial theorem in Section 3.3 was presented in connection with the problem of counting the number of k-element subsets of a set of n distinct elements. If we replace x by 1 and y by x in Theorem 3.3-5, we obtain the identity

$$(1 + x)^n = C(n, 0)x^0 + C(n, 1)x^1 + C(n, 2)x^2 + \cdots + C(n, n)x^n.$$

This identity expresses the function $g(x) = (1 + x)^n$ as a sum of nonnegative integral powers of x, each multiplied by a numerical coefficient. In particular, for $k = 0, 1, \ldots, n$, the coefficient of x^k is $C(n, k)$, which is the number of ways of selecting k objects from a set of n objects. The finite power series

$$g(x) = C(n, 0)x^0 + C(n, 1)x^1 + \cdots + C(n, n)x^n$$

is called the *generating function* for the combinatorial problem of counting the number of k-element subsets of an n-element set.

To see how these particular coefficients evolve in the algebraic expansion of $(1 + x)^n$, we illustrate with the value $n = 4$ and consider the product

$$(1 + x)(1 + x)(1 + x)(1 + x)$$

of four identical binomial terms. Each term of the expansion of this product is the product of four symbols, each symbol being either 1 or x and chosen from one of the four identical terms $1 + x$. For $k = 2$ we obtain the following six quadruple products.

1 1 x x	1 x 1 x	1 x x 1
x 1 1 x	x 1 x 1	x x 1 1

The product of the symbols in each set is 1^2x^2, and the sum of all x^2 terms is $C(4, 2)x^2$.

To fit this example of a generating function to the more customary model that we will use, we define

$$C(n, k) = 0 \qquad \text{for all } k > n.$$

This definition is sensible because $C(n, k)$ can still be interpreted (even when $n < k$) as the number of ways of choosing k objects from a set of n objects. With that convention we can write $g(x)$ as a power series

$$g(x) = C(n, 0)x^0 + C(n, 1)x^1 + C(n, 2)x^2 + \cdots = \sum_{k=0}^{\infty} C(n, k)x^k,$$

in which the coefficient of x^k is 0 for all $k > n$.

As a second example of a generating function, we recall the Fibonacci sequence described in Exercise 12 of Section 1.10:

$$1, 1, 2, 3, 5, 8, 13, 21, \ldots,$$

where each term after the second is the sum of the preceding two terms. Formally, we can define the kth Fibonacci number F_k recursively as follows:

$$F_0 = 1, \quad F_1 = 1, \quad F_{k+1} = F_k + F_{k-1} \quad \text{for all } k \geq 1.$$

(Observe that we are now denoting the first Fibonacci number as F_0 instead of as F_1.)

The generating function $g(x)$ for this number sequence is

$$g(x) = F_0 x^0 + F_1 x^1 + F_2 x^2 + \cdots + F_k x^k + \cdots.$$

The combinatorial problem associated with this sequence was published in 1202 by Fibonacci (Leonardo of Pisa):

Assume that each pair of rabbits produces a new pair of rabbits each month, starting after age 2 months. If all pairs survive, how many pairs of rabbits are living n months after a pair of newborn rabbits is obtained?

Observe that the number F_k is positive for each natural number k, whereas the number $C(n, k)$ is positive for only a finite set of values of k. In each case, however, the associated generating function can be expressed in the form of an infinite power series $\sum_{k=0}^{\infty} a_k x^k$.

It is important for us to realize that the concept of a generating function provides a significant generalization of the combinatorial techniques discussed earlier in this chapter, where each of several types of counting problems typically was analyzed by a technique tailored to that type of problem. In each case the analysis led to a formula that expressed the solution in terms of numerical values of the parameters of that problem. If we fix all but one of those parameters in a given problem, the formula defines a numerical sequence $\{a_p\}$ (finite or infinite) having the remaining parameter p as its variable: $a_0, a_1, a_2, \ldots, a_p, \ldots$. Since any finite sequence can be regarded as an infinite sequence in which all except a finite number of the terms are zero, we can associate with each such sequence a *formal power series* $\sum_{k=0}^{\infty} a_k x^k$ in which the coefficient a_k of x^k is the solution of that combinatorial problem when $p = k$.

The binomial theorem and the Fibonacci sequence suggest this generalization.

> **Definition** Let $a_0, a_1, a_2, \ldots, a_n, \ldots$ be a sequence of numbers. The *generating function* of that sequence is the formal power series
>
> $$g(x) = a_0 + a_1 x + a_2 x^2 + \cdots + a_n x^n + \cdots.$$

The term "formal power series" refers to any expression of the form

$$a_0 + a_1 x + a_2 x^2 + \cdots + a_n x^n + \cdots,$$

where each a_i is a number but the symbol x is *uninterpreted*. A finite formal power series has the form of a polynomial, but it is different because in a polynomial the symbol x is usually interpreted as a numerical variable (an unspecified number).

To use generating functions as a counting technique, we must first learn how to write a generating function $g(x)$ for which the coefficient of x^k equals the number that counts the desired set that corresponds to the integer k. Second, we must learn useful ways to rewrite $g(x)$ in a form that facilitates the numerical evaluation of the coefficient of x^k for each value of k. The rest of this section consists of illustrations of methods for constructing generating functions for a variety of combinatorial questions.

EXAMPLE 1   A person with a large supply of 2-, 3-, and 5-cent postage stamps wants to mail a letter that requires 37 cents postage. In how many ways can the exact postage be created with these stamps?

We use a method of analysis that is a modified form of "picture writing," advocated by George Pólya for setting up generating functions for various combinatorial problems. We first list in a table the several choices available for the number of stamps of each type to be placed on the letter. A large supply of each denomination is available, as shown in Table 3.9-1.

For each type of stamp we write a formal power series in which the coefficient of x^n is 1 if n occurs in the associated list of possible values, and the coefficient is 0 if n does not occur in that list. The exponent n is the value of the postage corresponding to using that many stamps of that denomination. Thus the three formal power series for this example are

2-cent: $x^0 + x^2 + x^4 + x^6 + x^8 + \cdots$

3-cent: $x^0 + x^3 + x^6 + x^9 + x^{12} + \cdots$

5-cent: $x^0 + x^5 + x^{10} + x^{15} + x^{20} + \cdots$

Type			Number Available				
2-cent	0	1	2	3	4	5	. . .
3-cent	0	1	2	3	4	5	. . .
5-cent	0	1	2	3	4	5	. . .
			Possible Values				
2-cent	0	2	4	6	8	10	. . .
3-cent	0	3	6	9	12	15	. . .
5-cent	0	5	10	15	20	25	. . .

Table 3.9-1

Now look at a few particular ways of making a choice of stamps that total 37 cents, as shown in Table 3.9-2. Observe that the first choice corresponds to the terms x^2, x^{15}, x^{20} in the three formal power series, respectively. The second choice selects the terms x^6, x^{21}, x^{10}. Each such choice will contribute one term of the form x^{37} in the product of the three power series, so the coefficient of x^{37} will be the number of ways of choosing 2-, 3-, and 5-cent denominations totaling 37 cents. Thus the sought-for generating function is that product:

$$g(x) = (1 + x^2 + x^4 + \cdots)(1 + x^3 + x^6 + \cdots)$$
$$\times (1 + x^5 + x^{10} + \cdots).$$

The next section examines ways to compute the coefficient of x^{37} in the expanded form of $g(x)$. For now we will be content with observing that the expanded form will start as follows:

$$g(x) = 1 + x^2 + x^3 + x^4 + 2x^5 + 2x^6 + 2x^7 + 3x^8 + \cdots.$$

These coefficients indicate that the number of ways to create a value of n cents in 2-, 3-, and 5-cent stamps is

0 ways to obtain a value of 1 cent

1 way to obtain a value of each of 0, 2, 3, 4 cents

2 ways to obtain a value of each of 5, 6, 7 cents

3 ways to obtain a value of 8 cents, and so on

	No.	Value	No.	Value	No.	Value	No.	Value
2-cent	1	2	3	6	1	2	1	2
3-cent	5	15	7	21	10	30	0	0
5-cent	4	20	2	10	1	5	7	35

Table 3.9-2

As an exercise, we may show that the coefficient of x^{37} in the formal power series $g(x)$ is 29. It can be done by listing systematically all ordered triples of natural numbers (a, b, c) such that $2a + 3b + 5c = 37$. For convenience we start with the largest possible value of c (namely 7) and list all possible values of b and a. Then we repeat for $c = 6$, and so on. Clearly, this method of enumeration is tedious, and for large values of n we want more efficient techniques.   ■

Before considering further examples of generating functions for counting problems, let us review the procedures used in Example 1 to construct a generating function $g(x)$ in which, for each natural number k, the coefficient of x^k specifies the number of ways of producing a total value of k cents by combining 2-, 3-, and 5-cent postage stamps. Let a, b, and c, respectively, denote the number of stamps used of each denomination. The value of those stamps is $(2a + 3b + 5c)$ cents, and the number of different combinations of stamps with total value k equals the number of ordered triples (a, b, c) that satisfy the equation $2a + 3b + 5c = k$. With each denomination of stamp we associate a power series in which each coefficient is 1, and the exponents are the values in cents that can be attained by using stamps of that denomination.

2-cent stamps: $x^0 + x^2 + x^4 + \cdots$

3-cent stamps: $x^0 + x^3 + x^6 + \cdots$

5-cent stamps: $x^0 + x^5 + x^{10} + \cdots$

Form the product of these series to obtain the desired generating function

$$g(x) = (1 + x^2 + x^4 + \cdots)(1 + x^3 + x^6 + \cdots)$$
$$\times (1 + x^5 + x^{10} + \cdots).$$

In expanded form $g(x)$ is the sum of all products of three terms, one from each of the three factors. Thus the coefficient of x^k in the expanded form of $g(x)$ is simply the number of products whose exponents sum to k. Each term in the first factor is of the form x^{2a}, whereas the terms in the second and third factors are of the forms x^{3b} and x^{5c}; hence, the product $x^{2a}x^{3b}x^{5c}$ will be x^k if and only if $2a + 3b + 5c = k$, as claimed.

In writing the generating function for Example 1 as a product of three formal power series (one for each of the three denominations of stamps), we need not include powers of x larger than 37. Thus, each of the three series could be a finite polynomial instead of an infinite series. But in Section 3.10, we shall see that the use of an infinite series can be advantageous, because some power series can be written in a form that facilitates the task of evaluating the coefficient of x^k in $g(x)$.

EXAMPLE 2 A doctor prescribes two types of medicine for a patient — at least 3 but no more than 5 pills of medicine A per day and at most 1 pill of medicine B. How many different combinations of pills can the patient take and still follow the doctor's orders?

The possibilities for the numbers of pills are

Medicine A: 3, 4, 5
Medicine B: 0, 1

The corresponding power series are

Power series A: $x^3 + x^4 + x^5$
Power series B: $1 + x$

Thus the generating function is

$$g(x) = (x^3 + x^4 + x^5)(1 + x)$$
$$= x^3(1 + x + x^2)(1 + x)$$
$$= x^3 + 2x^4 + 2x^5 + x^6.$$

The number of pills the patient can take each day is 3, 4, 5, and 6. So we need to sum the coefficients of those powers of x in $g(x)$; there are $1 + 2 + 2 + 1 = 6$ different ways in which the patient can take medicine within the prescribed ranges each day. This problem, of course, can be solved easily by listing all possibilities. ■

EXAMPLE 3 A party giver purchased one 6-pack of each of 4 brands of diet soft drinks to offer the guests. Construct a generating function for the problem of determining the number of different combinations of k bottles of soft drinks that could be consumed at the party.

For each type of soft drink the number of bottles consumed is some integer from 0 to 6. So the factor of $g(x)$ for each of the 4 types is

$$1 + x + x^2 + x^3 + \cdots x^6.$$

Hence,

$$g(x) = (1 + x + x^2 + \cdots + x^6)^4.$$

The coefficient of x^k (for $0 \le k \le 24$) in the expansion of $g(x)$ is the number of different combinations of k bottles of soft drinks that can be consumed. If p_i is the number of bottles consumed of type i, then the coefficient of x^k in $g(x)$ is the number of solutions in natural numbers of the equation

$$p_1 + p_2 + p_3 + p_4 = k,$$

because each such solution corresponds to a choice of x^{p_i} from factor i ($i = 1, 2, 3, 4$) in expanding the factored form of $g(x)$ to the polynomial form. We will continue with a modified form of this example in the next section to demonstrate how evaluation of the coefficient of x^k in $g(x)$ often can be simplified by using an infinite series instead of a polynomial for each factor of $g(x)$. To do this here, we would replace the generating function $g(x)$ by its series extension

$$G(x) = (1 + x + x^2 + \cdots + x^n + \cdots)^4. \qquad \blacksquare$$

Exercises 3.9

1. In an office of five accountants one of them is going out to buy a Hershey bar and offers to buy one candy bar for each of the other four. A second person specifies a Hershey bar; the other three express no preference but accept the offer. The candy store has Hershey bars, Heath bars, and Milky Ways.

   (a) Write a polynomial generating functon $g(x)$ for the number of selections of five candy bars in accordance with the expressed preferences.
   (b) Determine the value of the coefficient of x^5 in $g(x)$.
   (c) State the meaning of the coefficient determined in **b**.

2. Calculate the value of the coefficient that answers the postage stamp problem of Example 1 by using the systematic listing technique described in the text.

3. A person at a bus station wants to get a dollar changed into nickels, dimes, and quarters. Showing your reasoning, construct a generating function $g(x)$ for the number of ways in which it can be done. Which coefficient of $g(x)$ provides the answer to this question? (You need not evaluate that coefficient.)

4. Explain how your answer to Exercise 1**a** would be changed if

   (a) the number of accountants in the office was four instead of five;
   (b) the number of types of candy bars available for purchase was four instead of three.

5. As a variation of Example 1, assume that a person has only four 2-cent stamps, three 3-cent stamps, and two 5-cent stamps. Let a_k denote the number of ways in which k-cents worth of postage can be formed by using those stamps.

   (a) Write in factored form a polynomial generating function for the sequence $a_0, a_1, \ldots, a_n, \ldots$.

   (b) Write that generating function in expanded form, and list all values of k for which $a_k = 0$.

   (c) Calculate the value of $\sum_{k=0}^{\infty} a_k$ and interpret the meaning of that number in terms of postage.

6. A bakery packs its daily production of doughnuts into cartons with 10 dozen doughnuts in each carton and delivers the cartons to 4 stores. Stores A and B each want at most 2 cartons, and stores C and D each accept as many as are delivered.

   (a) Write a generating function for the sequence of numbers a_k, the number of ways in which k cartons can be delivered to the 4 stores with each store receiving at least 1 carton but not more than it is willing to accept.

   (b) Evaluate a_9 and interpret its meaning.

7. An interior decorator decides to purchase chairs selected from three models, with at least two chairs of each model and with an even number of chairs of one particular model.

   (a) Write a generating function for the sequence of numbers a_k, the number of ways in which k chairs can be selected.

   (b) Evaluate a_{12} and interpret its meaning.

8. Repeat Exercise 7 with the additional requirement that an odd number of chairs be selected from the other two models. Also show that a_{12} equals the number of solutions in even positive integers of the equation $a + b + c = 4$.

9. Use a generating function to solve Exercise 10 of Section 3.4.

3.10 Computation with Generating Functions

PROBLEM   For the party described in Example 3 of Section 3.9, suppose that three 6-packs of each brand were purchased. At the end of the party, the party giver found that 17 bottles of soft drinks were consumed. If at least 1 bottle of each variety was consumed, how many different combinations of the 4 varieties could have been consumed?   ■

This question reveals the need of new methods for evaluating the coefficients of large powers of x in a generating function—ways that are more efficient than polynomial multiplication. Such methods are provided by special identities, each of which expresses a formal power series as a single term in algebraic form. Although the number of such identities is quite large, many combinatorial problems can be solved by the following four identities.

Some Algebraic Identities (n is a positive integer)

1. $1 + y + y^2 + \cdots + y^n = (1 - y^{n+1})(1 - y)^{-1}$
2. $1 + y + y^2 + \cdots + y^n + \cdots = (1 - y)^{-1}$
3. $(1 + y)^n = C(n, 0) + C(n, 1)y + C(n, 2)y^2 + \cdots + C(n, n)y^n$
4. $(1 - y)^{-n} = C(n - 1, 0) + C(n, 1)y + C(n + 1, 2)y^2 + \cdots + C(n + p - 1, p)y^p + \cdots.$

The first identity is a special case of the familiar factorization rule

$$a^{n+1} - b^{n+1} = (a - b)(a^n + a^{n-1}b + a^{n-2}b^2 + \cdots + b^n);$$

it can also be derived by dividing $1 - y^{n+1}$ by $1 - y$. The second identity can be verified in the same two ways; in this case neither division nor multiplication terminates after a finite number of terms. In the division process, however, the successive quotients of 1 divided by $1 - y$ are of the form y^m for all natural numbers m, which is what identity 2 asserts.

Identity 3 is the familiar binomial theorem. To verify 4, we first apply 2 to obtain

$$(1 - y)^{-n} = [(1 - y)^{-1}]^n = (1 + y + y^2 + \cdots + y^k + \cdots)^n,$$

and we seek to show that the coefficient of y^p in the expansion of the product of the n identical power series,

$$(1 + y + y^2 + \cdots + y^k + \cdots)$$
$$\cdots (1 + y + y^2 + \cdots + y^k + \cdots),$$

is $C(n + p - 1, p)$. Power series are multiplied in exactly the same way as polynomials; that is, y^p appears in the product as a result of choosing one term from each of the n power series and multiplying to obtain $y^{k_1}y^{k_2} \cdots y^{k_n}$, subject to the further requirement that

(*) $$k_1 + k_2 + \cdots + k_n = p.$$

Each different way of choosing $k_1, \ldots, k_n$ to satisfy $(*)$ yields the product y^p, so the coefficient of y^p is simply the number of distinct n-tuples of nonnegative integers satisfying $(*)$. Or if we think of each k_i as the sum of k_i ones, then the desired coefficient of y^p is the number of ways of assigning p identical objects to n distinct boxes, namely, $C(p + n - 1, p)$, by using p in place of n and by using n in place of m in either Theorem 3.6-1 or Theorem 3.6-2. Thus identity 4 is verified.

EXAMPLE 1 Now we can solve the problem stated at the beginning of this section. The information that at least 1 bottle of each of the 4 types of diet drinks was consumed means that the x^0 term in each of the 4 identical factors of the generating function is not needed. Hence, we now write

$$g(x) = (x + x^2 + \cdots + x^n + \cdots)^4$$
$$= [x(1 + x + x^2 + \cdots + x^n + \cdots)]^4$$
$$= x^4(1 - x)^{-4},$$

where identity 2 was used to obtain the preceding equality. Then we apply identity 4 with $y = x$ and $n = 4$:

$$g(x) = x^4[C(3, 0) + C(4, 1)x + \cdots + C(p + 3, p)x^p + \cdots].$$

We want to determine the coefficient of x^{17}, which is the number of different combinations of 17 drinks of 4 types with each type represented at least once. Thus $p = 13$, and the desired coefficient is $C(16, 13) = 560$. ∎

In the preceding example we could also have used the generating function

$$f(x) = (x + x^2 + x^3 + \cdots + x^{18})^4$$
$$= x^4(1 + x + x^2 + \cdots + x^{17})^4.$$

By expanding the fourth power of the multinomial term in $f(x)$, we could determine the desired coefficient of x^{13} in that term, but the prospect of that much algebra is not attractive. So, at least in this case, the problem is made easier by using a generating function involving infinitely many powers of x.

EXAMPLE 2 At a child's birthday party various prizes were awarded to the 6 children, with each child receiving at least 2 prizes but no child receiving more than 4 prizes. If there were 20 prizes awarded, in how many ways could the numbers of prizes be distributed among the 6 children?

A generating function for the distribution of 20 prizes among the 6 children is

$$g(x) = (x^2 + x^3 + x^4)^6 = [x^2(1 + x + x^2)]^6$$
$$= x^{12}(1 + x + x^2)^6 = x^{12}h(x).$$

Hence we want to evaluate the coefficient of the x^8 term in

$$h(x) = (1 + x + x^2)^6 = (1 + x + x^2)^6(1 - x)^6(1 - x)^{-6}$$
$$= (1 - x^3)^6(1 - x)^{-6}.$$

If we now use identity 3 with $n = 6$ and $y = -x^3$ and identity 4 with $n = 6$ and $y = x$, we obtain

$$h(x) = (1 - x^3)^6(1 - x)^{-6},$$

where

$$(1 - x^3)^6 = C(6, 0) - C(6, 1)x^3 + \cdots + C(6, 6)x^{18}$$

and

$$(1 - x)^{-6} = C(5, 0) + C(6, 1)x + \cdots + C(p + 5, p)x^p + \cdots.$$

In this product the coefficient of x^8 will be

$$C(6, 0)C(13, 8) - C(6, 1)C(10, 5) + C(6, 2)C(7, 2),$$

which turns out to be

$$(1)(1287) - (6)(252) + 15(21) = 90.$$

Another way to evaluate the coefficient of x^8 in $h(x)$ is to think directly about the coefficient of x^8 in the expansion of $(1 + x + x^2)^6$. There are six identical terms, from each of which we are to choose either x^0 or x^1 or x^2 in such a way that the sum of the six chosen exponents is 8. We consider different cases according to the number, k, of terms from which we choose the x^2 term. The only possible values of k are 2, 3, and 4.

Case 1. If $k = 2$, we can choose the x^2 term from two of six factors in $C(6, 2)$ ways, and then we must choose x^1 from each of the remaining four factors.

Case 2. If $k = 3$, we can choose x^2 from three of the six terms in $C(6, 3)$ ways, and then we must choose x^1 from two of the remaining three terms. The number of choices for this case, therefore, is $C(6, 3)C(3, 2)$.

Case 3. If $k = 4$, we can choose x^2 from four of six factors in $C(6, 4)$ ways, and then we must choose x^0 from each of the two remaining factors.

Hence, the coefficient of x^8 in $h(x)$ is

$$C(6, 2) + C(6, 3)C(3, 2) + C(6, 4) = 15 + 20(3) + 15 = 90,$$

confirming our previous calculation. ∎

EXAMPLE 3 Determine the coefficient of x^{32} in the generating function

$$g(x) = (x^{10} + x^{15} + x^{20} + x^{25})^2(1 + x + x^2 + \cdots + x^{15}).$$

First we rewrite $g(x)$:

$$\begin{aligned}
g(x) &= x^{20}(1 + x^5 + x^{10} + x^{15})^2(1 + x + x^2 + \cdots + x^{15}) \\
&= x^{20}(1 + x^5)^2(1 + x^{10})^2(1 + x + x^2 + \cdots + x^{15}) \\
&= x^{20}(1 + 2x^5 + x^{10})(1 + 2x^{10} + x^{20})(1 + x + \cdots + x^{15}).
\end{aligned}$$

We need to find the coefficient of x^{12} in the product of the last three factors, so we need not retain any power larger than 12 in that product, which can be written

$$(1 + 2x^5 + 3x^{10} + \cdots)(1 + x + x^2 + \cdots + x^{12} + \cdots).$$

The x^{12} term in this product will be obtained as $1x^{12} + 2x^5x^7 + 3x^{10}x^2$, so the desired coefficient is 6. ∎

Exercises 3.10

1. Determine a numerical solution for the number of selections in Exercise 1 of Section 3.9 by using identities 2 and 4 to evaluate an appropriate coefficient of a generating function for that problem.

2. For each of the following generating functions and the given value of k, determine the coefficient of x^k in the power series form of $g(x)$.

   (a) $g(x) = (x^3 + x^4 + x^5 + \cdots)^7$ and $k = 32$,

(b) $g(x) = (1 + x + x^2 + \cdots + x^8)(1 + x + x^2 + \cdots)^5$ and $k = 18$,

(c) $g(x) = (1 + x^2 + x^3 + x^4 + \cdots)^2$ and $k = 7$.

3. Four students took a mathematics test consisting of 6 problems. Collectively these students solved exactly 18 problems correctly, with each student getting at least 3 right, but no one getting all of them right. In how many ways could the individual scores have occurred? Write a generating function for the sequence of possible scores, and then evaluate a suitable coefficient. Check your answer by listing the possible scores of the 4 students.

4. The Broke Music Society has decided to appeal for funds to enlarge its concert series next year. Each donor is classified as a Friend ($5 donation), a Patron ($25), or a Benefactor ($100). On the first day of the fund drive, 23 persons were solicited, some of whom decided to give nothing, and only 2 of whom even considered giving more than $5. But $135 was collected on that day. How many different lists of names of Friends, Patrons, and Benefactors could result from the first day's collections?

5. A Girl Scout troop with 16 members is selling boxes of cookies for the annual fundraising drive. The oldest scout can sell 2 boxes, 5 boxes, or no boxes. Each of the remaining 15 children can sell 1 box or no boxes. In how many ways could the troop sell 13 identical boxes of cookies? Write a generating function for the number of ways to sell the 13 boxes, and then evaluate the appropriate coefficient.

6. Thirty-six identical purple hard-boiled eggs are scattered across a large lawn for children to find in the annual egg hunt. Six children set out on the hunt and find all the eggs. The child who won the previous hunt may collect at most six eggs, but each of the other children may collect any number. Use generating functions and evaluate a suitable coefficient to find the number of ways the eggs can be distributed in the baskets when the children return.

7. Use generating functions to determine the number of ways to select 15 pieces of fruit from 5 types of fruit, with at least 1 of each type included.

8. Use generating functions to solve Exercise 8 of Section 3.6.

Chapter 4

GRAPHS:
THEORY AND
APPLICATIONS

The earliest known work in graph theory was published in 1736 when Leonhard Euler offered an ingenious solution to a folk puzzle of that era. (See Section 4.4.) For 200 years the subject grew sporadically, even haphazardly, as an informal collection of concepts and methods that were devised to solve certain challenging puzzles and scientific questions, some of which were deceptively difficult to answer. But mathematicians and scientists typically are intrigued by a difficult problem and are stimulated not only to solve that problem but also to develop methods for analyzing more general problems that underlie the original question.

By 1850, the pace of investigation of graphical notions accelerated. In 1936, Dénes König published the first comprehensive book on graph theory, thereby

launching graph theory as a recognized branch of mathematics with broad applications to comtemporary problems.

   This chapter presents graphical concepts and methods developed in the investigation of several puzzles of the famous "classical" period of graph theory (1736 –
1936). For pedagogical reasons, however, we have rearranged such topics from the
historical chronology in which they arose.

4.1   Concepts and Terminology

Section 2.2 introduced the notion of graphs as a geometric representation
of a binary relation R on a set S. Each element of S is represented by a
point, called a vertex, and each ordered pair (a, b) in the subset R of $S \times S$
is represented by an edge, a directed line segment, or a curve beginning at
a and terminating at b and containing no other vertex. When R is symmetric, the line segment joining a and b is undirected. In the symmetric
case the resulting figure is called a *graph;* otherwise it is called a *directed
graph.* In this chapter we shall study graphs, deferring directed graphs
until Chapter 5.

   Of course, graphs can arise in many different contexts, not only as a
means of representing a binary relation on some set. Therefore we begin
by rephrasing, in terms that are context free, the definition of graph given
in Section 2.2.

> A *graph G* is a structure $G = \{V, E\}$ that consists of a nonempty set V of
> points (called *vertices*) and a set E of segments (called *edges*) such that
> each edge e contains exactly two vertices, one vertex at each endpoint
> of the segment.

Thus, each edge of G can be regarded as an *unordered* pair $\{v, w\}$ of
vertices that may be either distinct or identical. Alternatively, an edge can
be regarded as a segment of a line or a curve joining a vertex either to itself
(a *loop*) or to a different vertex. In some applications it is convenient to
extend this definition of a graph to permit more than one edge to connect
any pair of vertices; in such a case we use the term *multigraph.*

Definition 4.1-1 Let $G = \{V, E\}$ be a graph.

(a) Two vertices v and w that are connected by an edge e are said to be *incident* with e and *adjacent* to one another.

(b) Likewise, two edges e and f that are both incident with vertex v are said to be *adjacent* at v.

(c) A *walk* W between vertices v_0 and v_k is a finite, nonempty sequence of adjacent vertices and edges of the form

$$v_0, e_1, v_1, e_2, \ldots, v_{k-1}, e_k, v_k,$$

where v_{i-1} and v_i are incident with e_i. The *length* of W is k. (No assumptions are made that the vertices or edges of W are distinct. A walk describes an unrestricted physical stroll in which vertices and edges can be traversed repeatedly in any direction.)

(d) A *trail* T in G is a walk W in which the edges e_i and e_j are distinct whenever $i \neq j$. (No edge of T is traversed more than once.) A trail that returns to its starting vertex is called a *circuit*.

(e) A *path* P in G is a trail T in which the vertices v_i and v_j are distinct whenever $i \neq j$. (No edge in P is repeated, nor does P return to any vertex previously visited.)

(f) A *simple circuit* C in G is a walk W of the form

$$v_0, e_1, v_1, e_2, \ldots, v_{k-1}, e_k, v_0,$$

in which $e_i \neq e_j$ and $v_i \neq v_j$ whenever $i \neq j$. (A simple circuit is a walk that returns to its starting vertex without traversing any edge more than once and without passing through any intermediate vertex more than once.)

Observe that in the graph of Figure 4.1-1(a) there is no trail to v_5 from any vertex, whereas $\{v_1, v_2\}, \{v_2, v_3\}$ and $\{v_1, v_2\}, \{v_2, v_4\}, \{v_4, v_3\}$ are two paths from v_1 to v_3. In some graphs and in all multigraphs, paths are more clearly described by labeling the edges directly rather than by listing adjacent pairs of vertices.

We now use the concept of trail to define an equivalence relation on the set V of vertices of a graph G. We say that vertex u is *reachable from* vertex v if and only if either $u = v$ or there is a trail from v to u in G. As an exercise, you may show that if there is a trail from v to u, then there is a path from v to u.

The relation "reachable from" is a binary relation on the set V of vertices of G. That relation is reflexive, by definition, and symmetric,

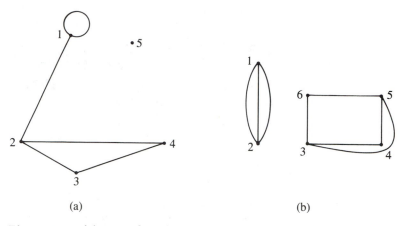

Figure 4.1-1 (a) A graph with one loop; (b) a multigraph with no loops.

because the edges of G are undirected. To show that it is transitive, we assume that u is reachable from w and that w is reachable from v. We want to find a trail from v to u. Although we can walk along a trail from v to w and then continue to walk along another trail from w to u, there is no assurance that the combined walk will be a trail, because the two given trails might contain a common edge, as shown in Figure 4.1-2.

Formally we write the two known trails in general form as

$$T_1: \{v, x_1\}, \{x_1, x_2\}, \ldots, \{x_i, a\}, \{a, b\}, \{b, x_{i+1}\}, \ldots, \{x_n, w\};$$
$$T_2: \{w, y_1\}, \{y_1, y_2\}, \ldots, \{y_j, c\}, \{c, d\}, \{d, y_{j+1}\}, \ldots, \{y_m, u\}.$$

Suppose that $\{a, b\}$ and $\{c, d\}$ represent the same edge of the graph; there is no loss of generality in assuming that $\{a, b\}$ is the first edge in T_1 that also occurs in T_2. We need to consider two cases, according to whether $a = c$ or $a = d$. First let $a = c$ and $b = d$. Then the trail T_3, defined by

$$T_3: \{v, x_1\}, \{x_1, x_2\}, \ldots, \{x_i, a\}, \{a, b\}, \{d, y_{j+1}\}, \ldots, \{y_m, u\},$$

proceeds along adjacent edges from v to u because $b = d$. It contains no edge more than once, because $\{a, b\}$ is the first edge of T_1 that occurs in T_2. As an exercise, you may complete the proof by considering the case in which $a = d$ and $b = c$. Hence "reachable from" is an equivalence relation on V, and the associated equivalence classes partition V. Each such equivalence class of vertices, together with the set of edges of G joining vertices of that class, is called a *component* of G. For example, the multigraph in Figure 4.1-1(b) has two components: $V_1 = \{1, 2\}$ and $V_2 = \{3, 4, 5, 6\}$, each with its associated edges. In that example V_1 is a multigraph, and V_2 is a graph (not a multigraph).

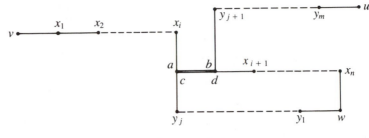

Figure 4.1-2

Definition 4.1-2 A graph G is *connected* if and only if G has exactly one component.

EXAMPLES OF GRAPHS

1. The definition of a graph requires that the set V of vertices be non-empty, but no such restriction is imposed on the set E of edges. If there are m vertices and no edges at all, then each vertex is a component. Such a structure is simply an m-element set; it is called an *empty* graph, even though its set of vertices is nonempty.

2. At the other extreme the graph with m vertices and one edge joining each pair of distinct vertices is called *the complete graph on m vertices* and is denoted by K_m. See Figure 4.1-3.

   Observe from Figure 4.1-3 that the vertices in a drawing of any graph need to be marked clearly, because it is sometimes necessary to draw nonincident edges in such a way that they cross on the plane and therefore *appear* to intersect. However, two edges of a graph have no points in common unless those edges meet at a vertex. Thus, it is useful to think of the edges as flexible strings, each being connected to two vertices in three-dimensional space before the entire system of vertices and strings is placed on a plane.

3. A *bipartite* graph G is a nonempty graph whose vertex set V is partitioned into two nonempty sets, L and R, where $V = L \cup R$ and $L \cap R = \emptyset$, such that *each* edge of G connects a vertex of L and a vertex of R. If there are m vertices in L and n vertices in R, and if each vertex in L is adjacent to each vertex in R, then G is called a *complete*

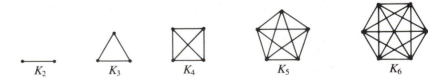

K_2 K_3 K_4 K_5 K_6

Figure 4.1-3 Complete graphs.

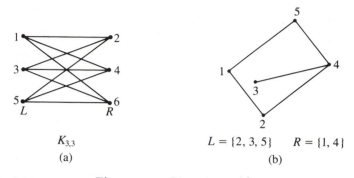

$K_{3,3}$

(a)

$L = \{2, 3, 5\}$ $R = \{1, 4\}$

(b)

Figure 4.1-4 Bipartite graphs.

bipartite graph and is denoted by $K_{m,n}$. In Figure 4.1-4 graphs (a) and (b) are bipartite, but only (a) is complete.   ■

A major challenge faced by a newcomer to graph theory is the formidable vocabulary that identifies the numerous concepts that distinguish one type of graph from another. This difficulty is largely unavoidable because graphs are general mathematical structures that can be used to model many different real-world problems. Therefore we must conscientiously study the vocabulary of graph theory and learn the meaning of each word.

We can simplify the terminology of graph theory by restricting our attention to special types of graphs. For example, a graph with no loops (that is, with no edge from a vertex to itself) is called a *simple* graph. Unless stated otherwise, we shall assume that each graph is simple. Similarly, we shall assume that each graph is *connected* (has only one component), *finite* (both V and E are finite sets), and *not a multigraph* (has at most one edge between each pair of distinct vertices).

Now, we still need one more concept. Let $G = \{V, E\}$ be a graph. The *complement, $C(G)$,* of G is the graph

$$C(G) = \{V, E'\},$$

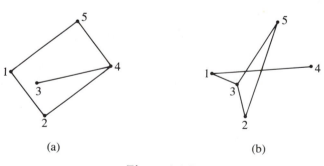

(a)   (b)

Figure 4.1-5

where the edge set E' consists of all pairs $\{x, y\}$ ($x, y \in V$ and $x \neq y$) such that $\{x, y\}$ is *not* an edge of G. For example, the graph in Figure 4.1-5(b) is the complement of the graph in Figure 4.1-5(a).

Exercises 4.1

1. **(a)** Draw K_4 and label its vertices A, B, C, and D in clockwise order.
   (b) Describe all paths from A to B.

2. **(a)** Let T denote any trail in a graph G from vertex v to vertex $u \neq v$. Describe how to use T to construct a path P from v to u such that each edge of P is an edge of T.
   (b) Let C be a circuit in G. Describe how to use C to construct a simple circuit C' such that each edge of C' is an edge of C.

3. **(a)** For $n = 2, 3, \ldots, 6$ make a table that lists the number of edges in K_n.
   (b) Study your list to see if you can correctly predict the number of edges in K_7. Check your prediction by counting.
   (c) State a formula for the number of edges of K_n.
   (d) Give reasons why your formula is correct. (That is, prove it.)

4. Show that the 8 corners and 12 edges of a cube form a bipartite graph. Is that graph a complete bipartite graph? Give reasons for your answer.

5. **(a)** Draw the complement of $K_{3,3}$.
   (b) Explain clearly why the complement of any complete bipartite graph has two components.

6. The following graph represents village streets. The postal department decides to place mailboxes at various street corners so that anyone in the village can mail a letter without walking beyond a corner adjacent to his or her residence. Assume that there are residences between each pair of adjacent corners.

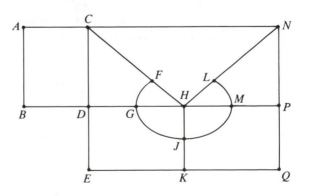

(a) Determine the smallest number of mailboxes needed, and show where each should be placed.

(b) Write the assertions that you need to establish to verify that your answer to **a** is correct.

7. Complete the proof in the text that "reachable from" is a transitive relation, by making an argument for the case in which $a = d$ and $b = c$ in Figure 4.1-2.

8. Prove the following assertions for every connected graph.

(a) If $m > 1$, any path that has m vertices has $m - 1$ edges.

(b) If $m > 2$, any simple circuit that has m distinct vertices has m edges.

4.2   Recognizing Different Graphs

The first question that we consider from the classical period of graph theory provides a good illustration of the cross-fertilization of ideas that frequently occurs between mathematics and other sciences—in this case, organic chemistry. In 1864, the Scottish chemist Alexander Crum-Brown proposed an improved graphical notation for representing the way in which atoms are bonded together to form molecules. To keep matters simple, we restrict our attention to the family of organic molecules that contain n atoms of carbon and $2n + 2$ atoms of hydrogen. Each hydrogen atom (H) is bonded in the molecule by a single bond, denoted by a connecting line segment. Each carbon atom (C) is bonded to other atoms in the molecule by four bonds. Such a molecule is written C_nH_{2n+2} and is a member of the paraffin family of compounds. For $n = 1$, 2, and 3 the chemical formulas are CH_4 (methane), C_2H_6 (ethane), and C_3H_8 (propane); the corresponding Crum-Brown representations (called *chemicographs*) are shown in Figure 4.2-1.

Because each hydrogen atom has a bonding valence of 1 and each carbon atom has a bonding valence of 4, the form of each of their chemi-

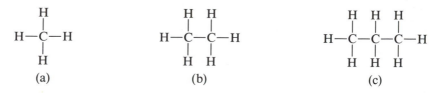

Figure 4.2-1 (a) Methane; (b) ethane; (c) propane.

cographs is uniquely determined. However, for $n = 4$ the paraffin compound C_4H_{10} can be represented by exactly two different Crum-Brown diagrams, as shown in Figure 4.2-2.

Observe carefully that each carbon atom in butane is bonded to either two or three hydrogen atoms, whereas in isobutane one of the carbon atoms is bonded to a single hydrogen atom. Chemical compounds, such as butane and isobutane, that have identical atomic constituents but different bonding configurations are called *isomers,* and they exhibit different physical properties. Thus a specific example of a general class of questions that occupied chemists in the mid-nineteenth century can be stated as follows: How many paraffin isomers C_nH_{2n+2} have n carbon atoms?

During the same period the English mathematician Arthur Cayley was developing a "theory of analytic forms called *trees.*" Cayley chose the name "tree" because the geometric diagrams that he used to represent his "analytic forms" reminded him of a tree, which today can be described as a connected graph that has no circuit. Cayley and his friend James J. Sylvester recognized a similarity between their studies and those of the chemists, and 20 years later Sylvester introduced the word "graph" to identify a more general form of the mathematical structure that he and Cayley had been investigating.

In a chemicograph an atom resides at each vertex, and a single bond between two atoms is simply an edge of the mathematical graph. Thus the graph of any paraffin molecule C_nH_{2n+2} has $3n + 2$ vertices (atoms) and $3n + 1$ edges (bonds). The latter number is obtained by totaling the valences of all the atoms and dividing by 2, because each bond is counted twice (once for each of the two atoms it connects). Hence, the graph of a paraffin molecule has no closed trails; no trail ever returns to a vertex that has been visited previously. (As an exercise in inductive proofs, we may prove that no graph with m vertices and $m - 1$ edges has a circuit.) This structure is of the type to which Cayley had assigned the name *tree;* we shall study trees in more detail in Section 4.7.

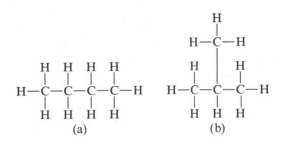

Figure 4.2-2 (a) Butane; (b) isobutane.

The problem undertaken now is a question about graphs that is closely related to this question about paraffin.

PROBLEM   How can we tell whether two graphs differ in some essential way?   ■

The problem is that for a given graph G we can draw a picture of G in many ways. How do we know whether two pictures, different in appearance, might both represent G? See Figure 4.2-3.

To answer such questions, we return to the definition of a graph as a set V together with a set E of unordered pairs of elements of V. Suppose we make a list of the elements of V, numbered in any order from 1 to n. We can then make a list of the elements of E as number pairs $\{i, j\}$; suppose there are m such pairs. Now to say that a given drawing D (n dots with certain pairs of dots connected by segments) *represents* G is to assert that D has n dots (numbered from 1 to n) and m segments that connect dot i and dot j if and only if the number pair $\{i, j\}$ or $\{j, i\}$ appears in the listing of E. This statement simply means that the vertices of G and the dots of D can be put into one-to-one correspondence in such a way that the list of number pairs of the edges of G and the list of number pairs of the endpoints of the segments of D are identical.

If two drawings represent G, then a one-to-one correspondence of the type described exists between G and each drawing, and those correspondences can be used to establish directly a one-to-one correspondence between the dots of the two drawings so that dot i and dot j are connected in one drawing if and only if the corresponding pair of dots are connected in the other drawing.

Definition 4.2-1   Two graphs $G(V, E)$ and $G'(V', E')$ are *isomorphic* when and only when a one-to-one function f with domain V and range V' exists such that for all u and v in V there is an edge between u and v in G if and only if there is an edge between $f(u)$ and $f(v)$ in G'.

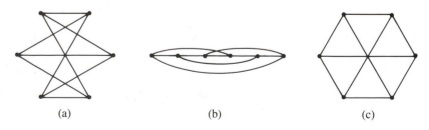

(a)   (b)   (c)

Figure 4.2-3 *Each drawing represents* $K_{3,3}$.

Finding differences between two drawings of graphs is often easier than proving that the graphs are isomorphic. If one has either more vertices or more edges than the other, quite clearly the graphs are not isomorphic. If the two vertex sets are the same size and the two edge sets are the same size, then we have to look for differences in the distribution of edges among pairs of vertices. If G has a vertex v with, say, five edges incident at v, but no vertex of G' has five incident edges, then G and G' are intrinsically different—that is, they are not isomorphic.

Definition 4.2-2 The *degree* (or valence) of a vertex v in a graph G is the number of edges of G that are incident at v. A *regular* graph is one in which each vertex has the same degree.

For example, each vertex in Figure 4.2-3 has degree 3.

Another technique for settling the isomorphism question is the following. If a graph G has many edges, then its complement $C(G)$ has only a few edges, because the complement of a graph with n vertices can be obtained by removing from the complete graph K_n all of the edges of G. For the same reason it follows that two graphs with n vertices are isomorphic if and only if the two complements are isomorphic.

If all attempts fail to prove that G and G' are not isomorphic, then the isomorphism question remains unsettled and we must look for a one-to-one correspondence of the type that will prove that G and G' are isomorphic.

EXAMPLE 1 Are the two graphs of Figure 4.2-4 isomorphic?

The initial appearance might tempt us to say that they are not isomorphic. But each has 7 vertices, and each vertex has degree 4 in each graph. Thus each graph has $7(4)/2 = 14$ edges. Let's try to define a one-to-one correspondence that will also be an isomorphism. Number the vertex set of G arbitrarily, and assign label $1'$ to any vertex of G', as in Figure 4.2-4.

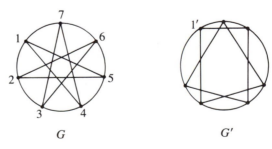

Figure 4.2-4

Now 1 is adjacent to 2, 4, 5, and 7 in G, so make a tentative assignment of those labels to vertices of G'. For example, try putting $2'$ in the same position as 2 is in G. Then $1'$ and $2'$ are both adjacent to the top position of G' and to the lower left position on G'. But in G each pair of adjacent numerals is adjacent to one and only one other vertex. Hence, in G' the positions corresponding to 2 and 7 in G must be filled by $4'$ and $5'$. We may verify that if we proceed counterclockwise around the circle of G', the labeling

$$1', 4', 7', 3', 6', 2', 5'$$

defines an isomorphism between G and G'.   ■

EXAMPLE 2   Consider the graphs in Figure 4.2-5. Each graph has ten vertices: two of degree 4, six of degree 3, and two of degree 2. The number of edges is $[2(4) + 6(3) + 2(2)]/2 = 15$. Observe, however, that the two vertices of degree 2 are adjacent in (a) but not in (b). Thus (a) and (b) are not isomorphic.   ■

We now justify this method of calculating the number of edges, and we relate that number to the number of vertices in a graph.

Theorem 4.2-1   Let G be a graph (finite and connected). The number of edges of G is one half of the sum of the degrees of all vertices in G.

Proof   Each edge connects two vertices and hence each is counted twice in adding together the degrees of all the vertices.   ■

Corollary   Any graph has an even number of vertices of odd degree.

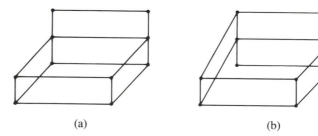

(a)   (b)

Figure 4.2-5

Proof From Theorem 4.2-1 the sum of the degrees of all vertices is even, and the sum of the degrees of all vertices of even degree is even. The difference between these two numbers is even, and it is also the sum of the degrees of all odd vertices. Thus the number of vertices of odd degree is even. ∎

Theorem 4.2-2 In any simple, connected graph G with v vertices the number e of edges satisfies the inequalities

$$v - 1 \le e \le \frac{v(v-1)}{2}$$

Proof The largest possible number of edges is the number of unordered pairs of vertices, so $e \le v(v-1)/2$, as claimed. To establish the inequality $v - 1 \le e$, we use the well-ordering property of N. Suppose that the inequality $v - 1 \le e$ is not true for some simple connected graph with v vertices and e edges. Then there is a smallest number s of vertices for which a simple, connected graph G exists, having s vertices and r edges, where $r < s - 1$. Furthermore, G can be chosen to be a graph having the smallest number r of edges such that $r < s - 1$. Then the deletion of any edge of G must separate G into two components, G_1 and G_2, collectively having s vertices and $r - 1$ edges. For $i = 1, 2$ let v_i and e_i denote the numbers of vertices and edges in G_i. Then

$$v_1 + v_2 = s, \quad \text{where } v_1, v_2 \text{ are positive,}$$
$$e_1 + e_2 = r - 1.$$

By our choice of s and G we know that

$$v_1 - 1 \le e_1, \quad v_2 - 1 \le e_2;$$

thus

$$s - 2 = (v_1 - 1) + (v_2 - 1) \le e_1 + e_2 = r - 1,$$

and

$$s - 1 \le r.$$

The last inequality contradicts the choice of r, so we reject the supposition that $v - 1 \le e$ fails to hold for some simple, connected graph. ∎

Exercises 4.2

1. Consider each of the six pairs of the following four graphs. In each case state a specific reason why those two graphs fail to be isomorphic.

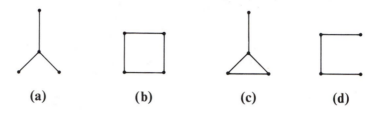

| (a) | (b) | (c) | (d) |

2. Label each of the three graphs in Figure 4.2-3 to establish an isomorphism with the graph $K_{3,3}$ as labeled in Figure 4.1-4(a).

3. Establish that the graphs of Figure 4.2-4 are isomorphic by drawing the complement of each and observing that each of the complements can be regarded as a simple circuit through seven vertices.

4. Determine which pairs of these graphs are isomorphic. Label isomorphic pairs to display an isomorphism.

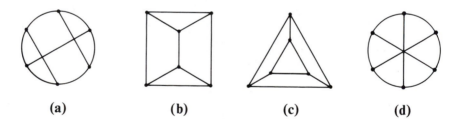

| (a) | (b) | (c) | (d) |

5. Determine which pairs of these graphs are isomorphic. Label isomorphic pairs to display an isomorphism.

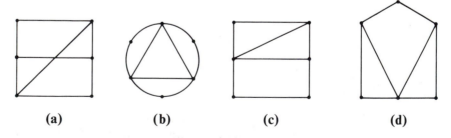

| (a) | (b) | (c) | (d) |

6. Determine which pairs of these graphs are isomorphic. Label isomorphic pairs to display each isomorphism.

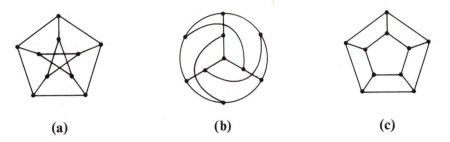

(a)   (b)   (c)

7. There are six nonisomorphic, connected graphs having four vertices. Draw a figure for each. Can you find any more such graphs?

8. Let u and v denote distinct vertices in a connected graph G.
   (a) Show that if a path from u to v has m edges, then it visits $m + 1$ vertices, including u and v.
   (b) Is the assertion in **a** valid if the word "path" is replaced by "trail"? Explain your reasoning.

9. Let C be a simple circuit in a connected graph G.
   (a) Show that the number of vertices in C equals the number of edges.
   (b) Is the assertion in **a** valid if the word "simple" is omitted? Explain your reasoning.

10. Determine the graphs of all of the paraffin molecules for which $n = 5$.

11. Determine the graphs of all of the alcohol molecules $C_nH_{2n+1}OH$ when $n = 4$. Regard OH as a single compound with valence 1.

4.3 Planar Graphs

PROBLEM   Which graphs can be drawn on the plane without any edges crossing except at vertices? ∎

   Attention was drawn to this question very gradually through a problem, apparently first published in 1917, although it was then regarded as old. Known as the "three utilities problem," it asks the reader to connect utility lines for water, gas, and electricity from three sites to each of three houses such that no utility lines cross. In terms of graphs, each vertex W, G, E must be connected on the plane to each vertex A, B, C with no edges crossing. See Figure 4.3-1. Of course, the graph will be the familiar bipartite graph $K_{3,3}$, and the question becomes: Can $K_{3,3}$ be drawn in the plane with no edges crossing (except at vertices)?

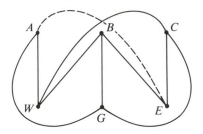

Figure 4.3-1 *Three utilities problem.*

About 75 years earlier the German mathematician A. F. Möbius had discussed another puzzle, not obviously related to the utility problem, that concerned a major landholder who had declared that upon his death his property was to be divided into five regions, one to be given to each of his five children. Each region was to share a boundary segment with each of the other regions. Can his wishes be carried out?

This problem can be translated into graph theory by representing each land parcel by a vertex and connecting a pair of vertices by an edge if and only if the two corresponding land areas share a common border. If the landholder's property can be divided as he desired, the resulting graph associated with the five parcels will be K_5. It can be proved that any map of an actual geographic region defines (by this representation scheme) a graph that can be drawn on the plane with no edges crossing. The Möbius problem thus becomes: Can K_5 be drawn in the plane with no edges crossing?

The studies stimulated by these brainteasers are highly important in the new technology of computers, robots, microchips, and printed circuits. Diagrams of electrical and electronic circuits look very much like graphs. In electricity the conducting paths are insulated wires in three-dimensional space, but in microelectronics they are analogous to bare wires on a plane and must not touch (except where intended at a junction). Hence, for a circuit to be reproduced in printed form, it must first be capable of being drawn in the plane with no edges crossing.

Definition 4.3-1 A graph G is *planar* if and only if it is isomorphic to a graph drawn in the plane with no pair of edges crossing.

Now the two puzzles are reduced to this form: Is $K_{3,3}$ or K_5 planar? The answer is no, as we soon become convinced if we repeatedly try to draw $K_{3,3}$ and K_5 in the plane.

To prove that neither graph is planar, we use a very simple formula, proved in 1752 by Euler, that applies to any planar graph (finite and connected).

Theorem 4.3-1 *(Euler)* Let G be a connected, planar graph having v vertices and e edges. Then any planar representation of G separates the plane into f nonoverlapping regions, where $f = e - v + 2$.

Before proving Euler's theorem, we illustrate the formula for the number of regions and restate the result in a more convenient form. A *planar region,* determined by a graph G, is any set F of points in the plane with the property that a curve can be drawn on the plane from any point of F to any other point of F without crossing any edge of G. We call each such region a *face* of G. If f represents the number of faces of G, then Euler's theorem asserts that for any planar representation of a planar graph

$$f - e + v = 2.$$

For this reason the number 2 is called the *Euler characteristic* of the plane.

The graph in Figure 4.3-2 determines four regions (don't forget the unbounded region of G). It has 10 edges and 8 vertices, and

$$f - e + v = 4 - 10 + 8 = 2,$$

as Euler claimed.

An edge $\{u, v\}$ of a planar graph G is called a *face-separating edge* if and only if each line segment that crosses $\{u, v\}$ contains points in more than one face of G. For example, in Figure 4.3-2 exactly 9 of the 10 edges are face-separating edges. Also observe that each finite face of a planar graph is bounded by a simple circuit.

Proof of Euler's Theorem We give an indirect proof that uses the well-ordering property of N. Suppose that the equation in Euler's theorem fails to hold for some connected planar graph. Then the set S of all values of e for which Euler's theorem fails for some connected planar graph with e edges is nonempty. By the well-ordering property

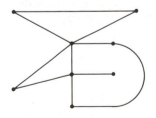

Figure 4.3-2

of N, S contains a smallest number m such that Euler's theorem is valid for all connected planar graphs for which $e < m$, but there is at least one graph G for which $e = m$ such that

$$f - e + v \neq 2.$$

Let $\{a, b\}$ denote any edge of G; we consider two cases, according to whether or not $\{a, b\}$ is a face-separating edge of G.

Case 1. If $\{a, b\}$ is a face-separating edge of G, let G_1 denote the graph obtained by deleting edge $\{a, b\}$ (but retaining vertices a and b in G_1). Then G_1 is planar because G is planar, and G_1 is connected because $\{a, b\}$ is one edge in some simple circuit of G. Let f_1, e_1, and v_1 denote the numbers of faces, edges, and vertices of G_1, respectively. Then $f_1 = f - 1$, $v_1 = v$, and $e_1 = e - 1 = m - 1$. Because $e_1 < m$, Euler's equation applies to G_1:

$$f_1 - e_1 + v_1 = 2,$$
$$(f - 1) - (e - 1) + v = 2,$$
$$f - e + v = 2,$$

which contradicts our selection of G.

Case 2. If $\{a, b\}$ is not a face-separating edge of G, the removal of edge $\{a, b\}$ separates G into two planar graphs G_2 and G_3, each of which is connected and has fewer than m edges. Because of the definition of m, the following equations must be valid:

$$f_2 - e_2 + v_2 = 2,$$
$$f_3 - e_3 + v_3 = 2,$$
$$(f_2 + f_3) - (e_2 + e_3) + (v_2 + v_3) = 4.$$

Observe that $v_2 + v_3 = v$, because G_2 and G_3 are disjoint graphs containing all vertices of G. And $e_2 + e_3 = e - 1$, because G_2 and G_3 are disjoint and contain all of the edges of G except $\{a, b\}$. Finally, $f_2 + f_3$ counts all of the faces of G, with the unbounded face counted twice. Hence,

$$(f + 1) - (e - 1 + v) = 4,$$
$$f - e + v = 2,$$

which again contradicts the selection of G.

If S is nonempty, Case 1 or Case 2 must be true; therefore, because each leads to a contradiction, we conclude that S must be empty, which completes the proof. ∎

For a planar graph that is not necessarily connected, a generalized form of Euler's theorem applies.

Theorem 4.3-2 Let G be a planar graph having c components, v vertices, e edges, and f faces. Then

$$f - e + v = c + 1.$$

Proof Label the components with subscripts $1, 2, 3, \ldots, c$. Then, for each subscript i, component i satisfies

$$f_i - e_i + v_i = 2.$$

For each i the number f_i is obtained by counting $f_i - 1$ bounded regions of the plane and one unbounded region of the plane. Ignoring temporarily the unbounded region of the plane, we let $b_i = f_i - 1$; the previous equation then becomes

$$b_i - e_i + v_i = 1 \qquad \text{for } i = 1, 2, \ldots, c,$$

where b_i is the number of finite regions of the plane bounded by edges of component i. If we sum v_i, e_i, and b_i over all values of i, we obtain the equations

$$
\begin{aligned}
v_1 + v_2 + \cdots + v_c &= v, \\
e_1 + e_2 + \cdots + e_c &= e, \\
b_1 + b_2 + \cdots + b_c &= f - 1,
\end{aligned}
$$

where each equation uses the fact that the components of G are disjoint, and the third equation uses the fact that f counts all finite regions of G and one infinite region of the plane. If we multiply each side of the second equation by -1 and then add the resulting three equations column by column, we obtain

$$1 + 1 + \cdots + 1 = v - e + f - 1.$$

Exactly c 1s are on the left side, so the equation can be rewritten in the desired form

$$c + 1 = f - e + v. \qquad \blacksquare$$

We now apply Euler's theorem to obtain some numerical relations that must hold in every connected planar graph G. We then use those relations to show that neither K_5 nor $K_{3,3}$ is planar. For such a G imagine walking around the perimeter of each face and counting each edge as we walk along it. Then each face-separating edge will be counted twice (once for each of the two faces that it separates). Because each face has three or more edges, the total edge count will be at least three times the number of faces. Thus,

$$3f \le \text{total edge count} \le 2e' \le 2e,$$

where e' denotes the number of face-separating edges of G. Therefore any connected planar graph satisfies the inequality

$$3f \le 2e.$$

Using this result with Euler's theorem, we obtain

$$3f - 3e + 3v = 6,$$
$$2e - 3e + 3v \ge 6,$$
$$(*) \qquad\qquad\qquad\qquad e \le 3v - 6$$

for any connected planar graph.

Note carefully that inequality $(*)$ must hold if a connected graph is planar. But the validity of $(*)$ does not guarantee that a connected graph is planar. That is, $(*)$ is a necessary condition for planarity, but it is not a sufficient condition.

Next consider K_5, which has $v = 5$ and $e = 10$, from which we conclude that K_5 is not planar, since $e = 10 > 9 = 3v - 6$.

Now suppose we consider a planar graph in which each face has at least four edges on its boundary. Then by the same reasoning we conclude that $4f \le 2e$, which combines with Euler's equation to yield $e \le 2v - 4$. As an exercise, we may fill in the details of this derivation and use it to prove that $K_{3,3}$ is not planar.

The roles of K_5 and $K_{3,3}$ in planarity are much more fundamental than those of simply being two examples of nonplanar graphs. In 1930, the Polish mathematician Kazamierz Kuratowski became the first to discover a necessary and sufficient condition for a graph to be nonplanar.

Kuratowski's Theorem   Any nonplanar graph contains a subgraph homeomorphic to $K_{3,3}$ or K_5.

Although the terms "subgraph" and "homeomorphic" are not defined until later, we shouldn't allow them to divert our attention from the principal meaning of the theorem, namely, that a graph G can be nonplanar in only two ways: Either G contains within itself a graph that is essentially the same as $K_{3,3}$, or else G contains within itself a graph that is essentially the same as K_5. Hence, $K_{3,3}$ and K_5, known for years in connection with recreational puzzles, assumed new significance because of Kuratowski's theorem.

Definition 4.3-2 Let $G_1 = \{V_1, E_1\}$ and $G_2 = \{V_2, E_2\}$ be graphs. G_2 is a *subgraph* of G_1 if and only if $V_2 \subseteq V_1$ and $E_2 \subseteq E_1$, and each edge in E_2 connects two vertices of V_2.

Note carefully that if G_2 is a subgraph of G_1 and $e = \{u, v\}$ is an edge of G_1, where u and v belong to V_2 and V_1, then e might or might not be an edge of G_2. See Figure 4.3-3.

Definition 4.3-3 Let $G_1 = \{V_1, E_1\}$ and let $e = \{u, v\} \in E_1$.

(a) Let a new vertex w of degree 2 be adjoined to V_1, located on e between u and v, so that e is replaced by two new edges, $\{u, w\}$ and $\{w, v\}$. The resulting graph G_2 is called a *simple subdivision* of G_1. In Figure 4.3-4, G_2 is a simple subdivision of G_3.

(b) A graph G_2 that results from a sequence of one or more simple subdivisions of G_1 is called a *subdivision* of G_1. In Figure 4.3-4, G_1 and G_2 are subdivisions of G_3.

(c) Two graphs, G_1 and G_2, are *homeomorphic* if and only if they are isomorphic or there exists a graph G_3 such that both G_1 and G_2 are subdivisions of G_3. In Figure 4.3-4, G_1 and G_2 are homeomorphic.

In the past half century graph theory has developed into a large area of mathematical inquiry, and Kuratowski's theorem has been supplemented by other planarity theorems—in particular, one by Hassler

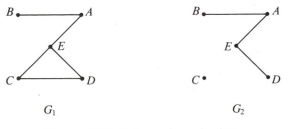

Figure 4.3-3 G_2 is a subgraph of G_1.

Whitney in 1932, given in terms of a concept of "duality" for graphs, and another, in 1937, by Saunders MacLane, given in terms of certain types of "basic circuits." Other characterizations of planarity have been stated more recently.

Although these planarity criteria are explicit, they are usually difficult to apply for determining whether a graph of some complexity does or does not contain within it a copy of K_5 or $K_{3,3}$. So even though the planarity problem was solved satisfactorily in the 1930s from a theoretical viewpoint, computational methods that answer the question in a reasonable amount of time were not developed until the 1960s and 1970s.

We conclude our concern with planarity by listing and illustrating some techniques that frequently enable us to resolve the issue of planarity, even though the methods might not succeed in all cases.

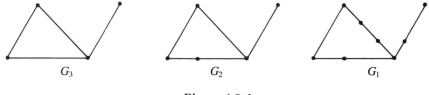

Figure 4.3-4

Suggestions for Investigating Planarity of G

Stage A. First simplify G by replacing it by a smaller graph G' that is known to be planar if and only if G is planar. Several types of simplification can be made.

1. G is planar if and only if each component of G is planar. Hence, this entire procedure can be applied to each component separately, one after another.

2. If the vertices of component G' can be separated into three non-empty subsets A, B, and C such that B contains only one vertex v, and if each path from a vertex in A to a vertex in C must pass through v, as in Figure 4.3-5, then G' is said to be 1-*connected* (or *separable*). If all vertices of A and all edges of $A \cup \{v\}$ are deleted, then the remaining vertices of C and edges of $C \cup \{v\}$ form a graph; a graph is also formed when the roles of A and C are reversed. If each of these graphs is planar, then so is G'. Hence, $A \cup \{v\}$ and $C \cup \{v\}$ can be treated separately as though each were a component.

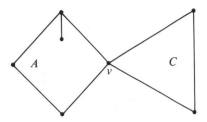

Figure 4.3-5 A separable graph.

3. If any nonseparable connected graph has multiple edges between two vertices, all but one of those edges can be deleted without affecting planarity. Therefore, we may delete any edge that joins two vertices that are also joined by another edge.

4. Any loop (an edge from v to v) can be deleted without affecting planarity.

5. Any vertex v of degree 2 and incident edges $\{u, v\}$ and $\{v, w\}$ can be removed and replaced by an edge $\{u, w\}$ without affecting planarity.

6. Steps 1 – 5 can be applied repeatedly until no further simplification is possible.

Stage B. For each simplified component identified in stage A, determine the number v of vertices and the number e of edges.

7. If $v < 5$, that component is planar. Why?

8. If $e < 9$, that component is planar. Why?

9. If $e > 3v - 6$, that component is nonplanar. Why?

10. If any simplified component is nonplanar, so is the original graph.

11. If each simplified component is planar, so is the original graph.

Stage C. At this point the planarity of G'' (the simplified component under consideration) might still be undecided.

12. When G'' is not prohibitively large, a reasonable procedure is to try to draw a figure for G'', starting with the longest circuit we can find. Then we add vertices and edges of G'' successively as long as possible with no edges crossing. If a planar figure for G'' is constructed in this way, we proceed to another component.

13. If we are unable to draw G'' in the plane without edges crossing, we try to find a $K_{3,3}$ or K_5 subgraph of G'' or to apply any other characterization of nonplanarity. When searching for a nonplanar subgraph of G'', it is convenient to ignore momentarily the presence of some vertices and edges in order to identify the 5 vertices and 10 edges of K_5 or the 6 vertices and 9 edges of $K_{3,3}$.

EXAMPLE   Figure 4.3-6 depicts a graph G_1 with two components; it has 15 vertices and 29 edges.

Beginning with stage A of the suggestions for studying planarity, we note in step 1 that G_1 has two components, consisting of vertices B through K and L through P. We first consider the component L through P because it has five vertices of degree 4, which is reminiscent of K_5. But upon closer examination, we see that that component is a multigraph; by step 3 we can eliminate one of each pair of double edges. We are then left with a 5-pointed star, which forms a simple circuit $LNPMOL$ that can be redrawn in planar form.

Now we consider the larger component of G_1, B through K, which is connected but separable, with H being a vertex of separation. So we may regard the graphs B through H and H through K separately, as though they were components. Clearly H through K is the complete graph on 4 vertices. By redrawing edge IK, we can eliminate one crossing in Figure 4.3-6, so that part of G_1 is also planar.

Finally we come to the graph G_2 (vertices B through H). It is planar if and only if G_1 is planar, because subgraphs H through K and L through P are. Following the steps of stage A, we note that G_2 is connected, nonseparable, and has no loops. After we remove one of the parallel GH edges, vertex G is of degree 2 and, by step 5, may be removed, leaving a simplified graph G_3 with four vertices of degree 4 and two of degree 3 (H and D). See Figure 4.3-7(a). Hence, G_3 cannot contain K_5, because K_5 has five vertices of degree 4.

No step of stage B is decisive, so we proceed to stage C and try to redraw G_3 in the plane, starting with a large simple circuit such as $BCDEFHB$,

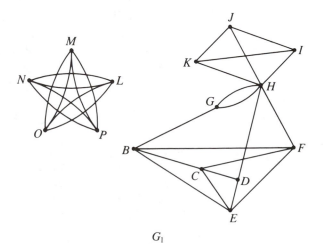

G_1

Figure 4.3-6

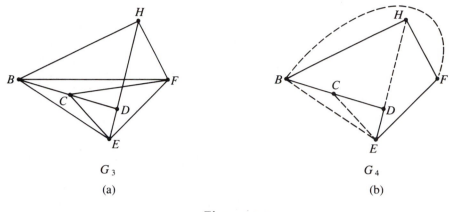

G_3

(a)

G_4

(b)

Figure 4.3-7

shown by solid lines in Figure 4.3-7(b). Then edges BE, CE, DH, and BF can be added in the plane with no edges crossing, producing a planar subgraph G_4 of G_3 with CF as the only missing edge of G_3. But there is no clear way to adjoin CF to G_4 without violating planarity. It is conceivable that a different method of redrawing G_3 might succeed in finding a planar representation of G_3. However, if we tend to doubt that G_3 is planar, we should try to locate $K_{3,3}$ within G_3. Because G_3 has 6 vertices and 11 edges, a model of $K_{3,3}$ within G_3 must contain every vertex of G_3 but only 9 of the 11 edges. Observe that in G_3

B is adjacent to C, E, F, H.

D is adjacent to C, E, H.

F is adjacent to B, C, E, H.

If we ignore edges BF and CE in G_3, each of the vertices $\{B, D, F\}$ is connected to each of the vertices $\{C, E, H\}$, forming a model of $K_{3,3}$ as a subgraph of G_3 and hence within G_2. See Figure 4.3-8. Thus G_1 is non-planar.

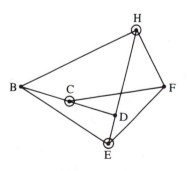

Figure 4.3-8

Exercises 4.3

1. Try to solve the landholder puzzle by attempting to draw five regions on the plane so that each is an across-the-fence neighbor of the other four. Show and explain in words the outcome of your attempts.

2. Try to solve the landholder puzzle by drawing the regions on the surface of a doughnut or a lifesaver or an inflated inner tube (or on a drawing of any of those objects).

3. Show that any connected planar graph has a vertex of degree less than 6.

4. Fill in the details of deriving the inequality $e \leq 2v - 4$ that must hold in a planar graph if each face has four or more edges along its boundary.

5. Show in detail how the inequality of Exercise 4 can be used to prove that $K_{3,3}$ is not planar.

6. Derive an inequality relating e and v that must hold in a planar graph if each face has five or more edges.

7. Answer each of the "why" questions raised in the description of stage B of the procedures for testing planarity.

8. Explain how you can be sure that K_n is planar for all positive $n < 5$ but nonplanar for all $n \geq 5$.

9. Determine whether or not each of the following graphs is planar. Show each step of your method.

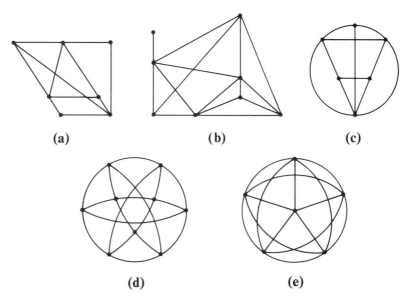

(a)   (b)   (c)

(d)   (e)

10. A Platonic solid S_n is a regular polyhedron—that is, a three-dimensional solid, each face of which is a regular plane polygon having $n \geq 3$ edges of equal length. Thus, all faces of S_n are congruent, all edges are congruent, and all face angles are congruent of size $180°(n-2)/n$. Each face has n vertices, and each vertex has m incident edges, where $m \geq 3$.

(a) By considering the sum of all the face angles at any vertex (which must be less than 360°), derive the inequality

$$3 \leq m < 6.$$

(b) By counting the total number of edges in two ways, obtain the equations

$$\frac{mv}{2} = e = \frac{nf}{2}, \qquad mv = 2e = nf.$$

(c) Use **a** and **b** and Euler's equation to determine the essential data for each of the Platonic solids by filling in the rest of the following table.

m	n	f	e	v	Shape of face
3	3				Triangle
:					
:					

4.4 Euler Trails and Circuits

In a city once called Königsberg the Pregel River flowed around two islands that were connected to each bank of the river and to each other by seven bridges, as shown in Figure 4.4-1. Someone wondered if it were possible to start from home, walk across each bridge exactly once, and return home (without recrossing any bridge). The Königsberg question reached Euler at the academy in St. Petersburg, and by 1736 he published a solution of a more general puzzle. That was the first article on graph theory, and it was notable not only for its results but also for Euler's insight in representing each land area by a letter and each passage across a bridge by the pair of letters of the land areas exited and entered, which resulted in an algebraic representation of the graph.

Let's label Figure 4.4-1 with A for the north shore, B for the western island, C for the south shore, and D for the eastern island; next imagine

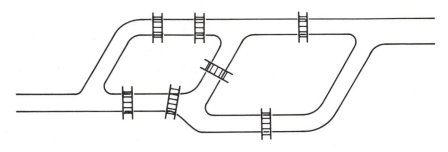

Figure 4.4-1 *Bridges of Königsberg.*

shrinking each land area to a point, with the bridges as the only connecting links (edges) between points (vertices). Then we have the multigraph of Figure 4.4-2. The problem is to start at any vertex, travel over each edge exactly once, and end at the starting vertex. Such a route in a graph was defined in Section 4.1 to be a circuit, whether or not each edge of the graph was covered.

PROBLEM   For which graphs is there a circuit that passes along each edge exactly once?   ■

> ***Definition 4.4-1*** Let G be a graph or multigraph. An *Euler trail* of G is a trail that covers each edge exactly once. An *Euler circuit* is an Euler trail that ends at its starting vertex.

You may remember some children's puzzle books in which one puzzle asked you to connect the numbered dots in order, and the picture turned out to be a clown or a chicken. Another puzzle asked, "Can you draw this picture without lifting your pencil or retracing any lines?" These were thinly disguised examples of Euler trails and circuits.

We are considering only connected graphs, so an Euler trail must visit each vertex of G at least once in traversing each edge. Let u be the starting vertex, w the ending vertex, and v any vertex different from u and w. Each time the trail arrives at v it must continue on, so the number of edges at v is

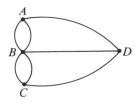

Figure 4.4-2

twice the number of times that the trail visits v. Thus the degree of each vertex (except u and w) must be even, if there is an Euler trail from u to w. Now suppose $u \neq w$, so the trail is not a circuit. The degree of u must be odd because the trail leaves u on one edge and each revisit to u covers two edges — one upon arrival and one upon departure. A similar count for w shows that w must be of odd degree. But if u and w coincide, the existence of an Euler circuit in G implies that u, and every vertex, must be of even degree.

Theorem 4.4-1 A connected graph or multigraph G has an Euler circuit if and only if each vertex of G has even degree. G has an Euler trail from u to w ($w \neq u$) if and only if u and w each has odd degree and every other vertex has even degree.

Proof The argument preceding the theorem shows that the statements about vertex degrees are necessary conditions for the existence of an Euler circuit or Euler trail, respectively. To prove the converse, we describe a method that can be used to construct an Euler circuit or trail when the corresponding conditions are satisfied.

We first prove the statement about Euler trails. Suppose that each vertex of G has even degree, except that u and w each has odd degree. We shall construct an Euler trail from u to w. Start at u along any edge extending from u and form as long a chain of vertices as possible, marking each edge (with numbers or arrows, for example) as it is traversed. After the trail leaves u it can continue through each vertex that it reaches (with the possible exception of w) because there will always be an even number of unmarked edges. Hence, if the trail is forced to stop, it must have reached w, and no further edges are available for exit from w. At this stage it is quite possible that some edges are still unmarked, as indicated by the dashed edges in Figure 4.4-3.

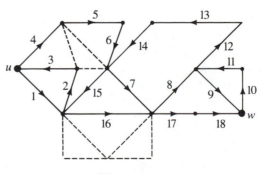

Figure 4.4-3

If we regard the solid edges as having been erased, the dashed edges connect vertices, each of which is of even degree, and those edges and vertices form a graph G', conceivably with several components. Because the original graph G was connected, each dashed component has at least one vertex in the solid trail. Start at any such vertex v' and trace a trail along dashed edges in that component. Because each vertex has even degree, the trail is forced to stop in that component only when it returns to v', and all the edges at v' are now marked. Thus a circuit has been traced in that component. Since the first trail passed through v', we have sequences of vertices as follows:

First (trail): $u, v_1, v_2, \ldots, v', \ldots, w$
Second (circuit): $v', \ldots, v'.$

If we break the first trail at v', insert the second sequence, and continue on from v' along the first trail, we have a new and larger trail from u to w.

Now continue this process with each component of dashed edges. At each stage we can fit newly marked edges into the previous trail to obtain a longer trail. Because the number of edges of G is finite, the process will end eventually, but only after all edges have been traversed, thus producing an Euler trail.

A proof for the statement of the first part of this theorem (Euler circuit) is readily obtained from this proof of the second part. To demonstrate that an Euler circuit exists when each vertex has even degree, we choose any vertex v and let its degree be $2k$. We next choose any edge e at v and think of v as being two vertices at the same position: a vertex u of odd degree $2k - 1$ and a vertex w of odd degree 1 with e as its only edge. We now build an Euler trail from u along edge e and eventually to w. Because u and w coincide with v, we have constructed an Euler circuit starting and ending at v.   ∎

EXAMPLE   Remote control of equipment depends on accurate transmission of data between the equipment and the control mechanism. Such data reports some aspect of the current state of the equipment; the control mechanism receives that data, compares it with a desired state, and transmits a correction instruction (perhaps a zero correction) to the equipment. In many instances the transmission is a sequence of 0s and 1s that has been generated in the equipment to correspond to some physical condition within that equipment, such as direction, temperature, rate of rotation, and so on. The description of physical condition by numerical data requires an analog-to-digital conversion. In this example we illustrate how an Euler circuit is employed in one type of conversion device.

Imagine that the physical condition to be monitored is indicated in the equipment by the angular orientation of a dial attached to the end of a shaft that rotates as the condition changes. The rim of the face of the shaft is divided into electrically separate (insulated) sections, each of which carries a constant voltage of either 0 or 1. A set of n electric sensors is positioned to be constantly in contact with n consecutive sectors of the dial. The system is diagrammed in Figure 4.4-4 for $n = 3$.

For each position of the dial the three output terminals each carry the voltage of the sector with which it is in contact — a sequence of three digits, each digit being 0 or 1. Because there are only 8 different three-term sequences of two distinct symbols ($8 = 2^3$), the number of insulated sectors in Figure 4.4-4 is 8. In an n-contact device there would be 2^n sectors on the edge of the dial.

We would like to answer the following question:

Is it possible to assign voltages 0 or 1 to each sector in such a manner that as the dial rotates through $360°$ (say, clockwise) the 8 three-digit binary numbers appear in some sequence on the output terminals?

If we think of this question in terms of an Euler circuit, we want to devise a graph with 8 edges, each edge bearing a different label of the form ABC, where each letter is 0 or 1. Each 3-digit binary number can be generated by affixing either 0 or 1 on the right-hand side of a suitable 2-digit binary number, of which there are 4. Hence, we think of a graph with 4 vertices and 8 edges. Because we have specified a direction of rotation of the dial, we need to describe a directed graph. (See Section 2.2.)

Start with a vertex labeled 00. From it will extend two edges labeled 000 and 001 that lead to vertices 00 and 01, obtained by deleting the left-hand digit of the edge that leads into the vertex. See Figure 4.4-5, which shows the evolution of the graph in three stages.

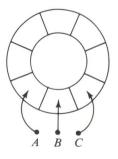

Figure 4.4-4

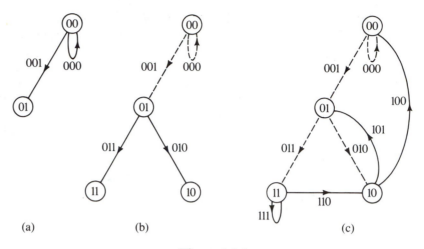

Figure 4.4-5

The directed graph in Figure 4.4-5(c) has an even number of edges incident to each vertex; moreover, at each vertex the number of edges leading in equals the number leading out. That is, at each vertex the *in-degree* equals the *out-degree,* which is a necessary and sufficient condition for a directed graph to have an Euler circuit. And an Euler circuit is easy to see: Starting at 00, follow edges 000, 001, 011, 111, 110, 101, 010, 100. These binary numbers represent the sequence of decimal numbers 0, 1, 3, 7, 6, 5, 2, 4. It also indicates that fixed voltages can be assigned to the sectors of the dial as shown in Figure 4.4-6.

It is reasonable to believe (and it is true) that this construction method applies to a dial with 2^n sectors for any $n > 0$ and that the precision of the information increases as n increases. Because electronic control devices typically process information in the form of sequences of binary digits, analog-to-digital conversion devices are used frequently in modern technology.

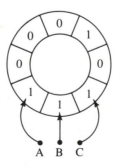

Figure 4.4-6   ■

Exercises 4.4

1. In each of the graphs determine whether there is an Euler circuit, an Euler trail but no Euler circuit, or neither. Justify each answer and list the vertex sequence of a trail or circuit if one exists.

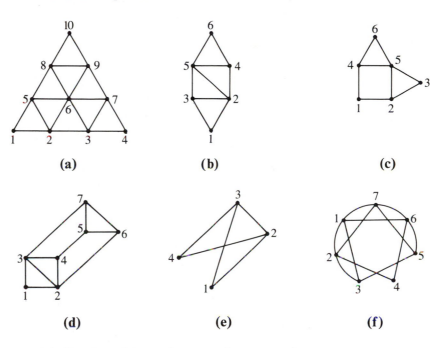

(a)   (b)   (c)

(d)   (e)   (f)

(g) Graph **a** with one bottom edge removed.
(h) Graph **a** with two bottom edges removed. Does it matter which two are removed?

2. Is it possible to trace all of the edges of a cube in one trail? Explain your reasoning.

3. State necessary and sufficient conditions for a separable graph to have an Euler circuit (as defined in Section 4.3). Explain your reasoning.

4. Prove that the complete graph K_n has an Euler circuit when n is odd and fails to have an Euler trail when n is even and $n \neq 2$.

5. Determine a sequence in which the 16 sectors of a dial could be labeled in order to construct an analog-to-digital converter when $n = 4$. Show your work, and write in decimal form the sequence of 16 numbers thus generated.

6. Suppose that a graph G has precisely $2k$ vertices of odd degree, where $k > 1$. (Recall from Section 4.2 that the number of vertices of odd

degree is even.) Show that by adding $k - 1$ new edges to G, you can obtain a graph or multigraph with an Euler trail. Also show that by adding k new edges to G, you can obtain a graph or a multigraph with an Euler circuit.

7. An independent sanitary engineer (garbage collector) can collect trash from each house in her city in five days, so she establishes five routes that cover all subscribers once a week. As a truck passes along a street, collections are made from subscribers on both sides of the street to avoid, whenever possible, traveling over any street more than once each week. What should the engineer take into account to plan five routes that will minimize the total mileage traveled in a week? Write a brief report (less than a page) that considers various kinds of street configurations and how they affect the planning. The previous exercise might help.

8. For each graph or multigraph describe an Euler trail, or explain why none exists. Also describe an Euler circuit or explain why none exists.

(a)

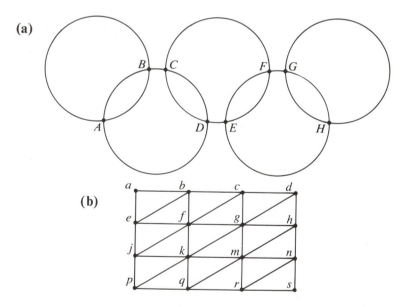

(b)

9. Determine the smallest number of bridges that would have to be closed in Königsberg to enable the citizens to stroll along an Euler circuit across the remaining bridges. Draw a graph or multigraph of the modified city, labeled as in Figure 4.4-2.

10. Determine the smallest number of new bridges in Königsberg that would enable the citizens to stroll along an Euler circuit across all of the bridges. Draw a graph or multigraph of the modified city, labeled as in Figure 4.4-2.

4.5 Hamilton Paths and Circuits

PROBLEM   For which graphs is there a simple circuit that passes through each vertex exactly once? ∎

    This problem is the same question (about vertices) that we asked (about edges) in the previous section. And only rarely in mathematics does there occur a pair of problems so superficially similar but so fundamentally different. Indeed, if you can solve this problem, your fame (if not your fortune) will be firmly established!

    This question is more than a century old. It was posed in 1855 by an English clergyman, Thomas Kirkman, in terms of simple circuits through each vertex of a given polyhedron. At the same time William R. Hamilton had been studying a variety of more specialized questions about vertex circuits on a dodecahedron. The dodecahedron, one of the five Platonic solids, has 12 faces, all of which are congruent regular pentagons, 30 edges, and 20 vertices each of degree 3. Hamilton had been trying to develop a method of representing rotations in three-dimensional space and had published his theory of quaternions in 1853. Apparently as an outgrowth of those investigations, he invented and marketed a board game that he called *the Icosian game,* in reference to the 20 vertices of the dodecahedron. Hamilton projected the dodecahedron onto the plane to obtain the graph in Figure 4.5-1. The Icosian game presented that graph on a wooden board with a circular indentation at each vertex, along with 20 numbered circular pieces, each of which could be placed in any vertex indentation to number the edges in various ways that would define a simple circuit along edges, passing once and only once through each vertex. A few of Hamilton's proposed games are described as exercises.

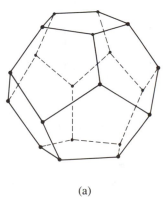

(a)

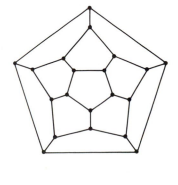

(b)

Figure 4.5-1 (a) Dodecahedron; (b) Icosian graph.

Although Kirkman seems to have a prior claim for the general question that introduced this section, history has firmly attached Hamilton's name to it.

Definition 4.5-1 Let G be a graph. A *Hamilton path* in G is a path that passes through each vertex exactly once. A *Hamilton circuit* in G is a simple circuit that passes through each vertex of G exactly once.

Observe that G must be connected if a Hamilton path exists, but connectedness does *not* guarantee that existence. For example, consider a connected graph with three edges and four vertices, one of degree 3. Next notice that each time any path goes through a vertex it uses exactly two edges incident to that vertex. Hence, if a path includes n vertices, each exactly once, then exactly $n - 1$ edges are covered, and if a simple circuit includes n vertices then exactly n edges are covered. Therefore it follows that if a vertex has degree 2, every Hamilton circuit must cover those two edges. The use of those edges then reduces the number of edge choices at both the previous vertex and the following vertex. Similarly, once a path passes through a vertex v, all other edges incident at v can be eliminated from consideration for use in any other segment of that path.

Through such reasoning, we can sometimes either construct a Hamilton circuit or prove that none exists.

EXAMPLE 1   Refer to the graph in Figure 4.5-2(a). We can start at any vertex to build a Hamilton circuit if one exists. Denote the center by A, proceed directly downward to B, and left to C to obtain (b). A return to A is premature, so we must proceed all the way around to I and then to A. That was too easy, so let's try again with another graph.

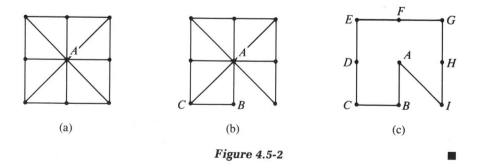

(a)   (b)   (c)

Figure 4.5-2   ■

EXAMPLE 2   In Figure 4.5-3 vertices A, C, E, G are of degree 2, so every side of each square must be an edge of any Hamilton path. To pass from one square to the other and back requires two more edges. But no Hamilton circuit on eight vertices can have more than eight edges. No Hamilton circuit exists.

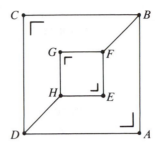

Figure 4.5-3

Although the Hamilton circuit problem appears to be quite similar to the Euler circuit problem, surprisingly, after a century of continued interest by mathematicians, no complete answer to the question posed at the start of this section has been given. Aside from intellectual interest, there are strong economic reasons to continue to investigate Hamilton paths and circuits. Companies that employ sales representatives to call on customers in a given area would gladly save travel costs and time by planning the most economical routes or circuits. Interstate transportation companies would like to use their equipment more efficiently and to obtain new routes (or perhaps drop some existing ones) to improve profitability. And the night watchman would like to know how to walk to all assigned punch-in stations without taking extra steps. We will return to such questions later in this chapter and in Chapter 5.

Even though a Hamilton circuit on a graph G with n vertices has only n edges, the number of edges of G is important. For example, for every $n > 1$, the complete graph K_n has a Hamilton path. For odd values of n one can say more.

For n odd $(n > 1)$ K_n has $(n-1)/2$ Hamilton circuits, no two of which have any edge in common.

This result reveals the effect of having enough options at each vertex when a Hamilton circuit is being sought. Another result is:

A graph G with n vertices $(n > 2)$ has a Hamilton circuit if each vertex has degree at least $n/2$.

This theorem assumes that the graph is connected, without loops, and without multiple edges. The condition is sufficient but not necessary. Notice, for example, that a graph with five vertices, each of degree 2, has a Hamilton circuit.

A third example of a sufficient (but not necessary) condition for a Hamilton circuit is stated in this theorem.

A graph G with n vertices $(n > 2)$ has a Hamilton circuit if for each pair u and w of nonadjacent vertices

$$\deg(u) + \deg(w) \geq n.$$

Next we consider a different type of puzzle that can be analyzed by means of Hamilton circuits.

EXAMPLE 3   A puzzle popular in the early 1970s was called Instant Insanity. There were four cubes, with each face of each cube colored either red, white, blue, or green. In Figure 4.5-4 we show a cut and flattened cube with each face numbered. The color of each face of each cube in a typical design of the puzzle is shown in the accompanying table. Notice that when the flattened cube is folded into a cube, faces 1 and 4 are opposite, as are 2 and 5, and 3 and 6. The object is to stack the cubes vertically into a tower in such a way that each color appears on each of the four vertical sides of the tower. For clarity we imagine that the cubes are stacked with one vertical side of the tower facing north. Observe that if a particular cube has face 1 headed north, then face 4 is headed south. Then exactly one of the following four orientations occurs for that cube:

3 east, 6 west, 2 up, 5 down

6 east, 3 west, 5 up, 2 down

5 east, 2 west, 3 up, 6 down

2 east, 5 west, 6 up, 3 down

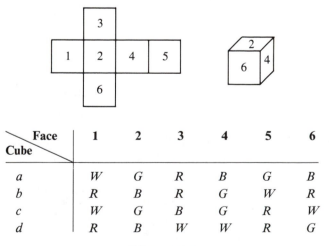

Face Cube	1	2	3	4	5	6
a	*W*	*G*	*R*	*B*	*G*	*B*
b	*R*	*B*	*R*	*G*	*W*	*R*
c	*W*	*G*	*B*	*G*	*R*	*W*
d	*R*	*B*	*W*	*W*	*R*	*G*

Figure 4.5-4

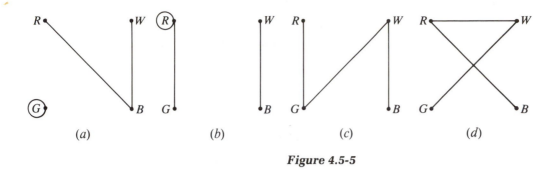

(a)   *(b)*   *(c)*   *(d)*

Figure 4.5-5

For each cube we draw a graph with four vertices (one vertex for each color) and three edges (one edge connecting the two colors of each pair of opposite faces of that cube). Thus there are four graphs, as in Figure 4.5-5.

For the next step it is convenient first to superimpose the four graphs to obtain one multigraph M with loops. We also affix a letter to each edge to record the cube to which that edge corresponds. See Figure 4.5-6. Observe that the degree of each vertex is the number of faces bearing that color among the four cubes of the puzzle, and each letter used as an edge label should appear three times.

Suppose there is a Hamilton circuit of M such that each edge bears a different label; for example, the edges a, b, d, c that are marked in Figure 4.5-6 describe the color cycle $RBWGR$. Adjacent vertices indicate opposite sides, so we can interpret this as meaning that the cubes can be stacked in the order a, b, d, c in such a way that one vertical face will bear the colors $RBWG$ and the opposite face will be $BWGR$. The order in which the cubes are stacked is irrelevant, of course, in solving the puzzle.

So now we have the north and south faces of the tower properly arranged, with each color showing once on each of those faces. Next we want

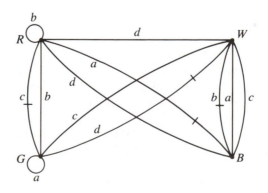

Figure 4.5-6 Multigraph M.

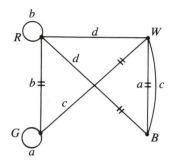

Figure 4.5-7

to ensure that the east–west faces also show four colors (without changing the north–south color combinations). Suppose there is a Hamilton circuit among the unmarked edges of M; then the puzzle has a solution. Figure 4.5-7 shows the graph of M with the previously marked edges removed and a new Hamilton circuit marked. Edges d, a, c, b trace out the color cycle $RBWGR$. (The coincidence of this cycle with the color cycle of the first Hamilton circuit is a peculiarity of this example.)

To see that the two Hamilton circuits indicated in Figures 4.5-6 and 4.5-7 describe a solution of this puzzle, we incorporate our results in Table 4.5-1.

	Figure 4.5-6		Figure 4.5-7	
Cube	**North**	**South**	**East**	**West**
a	R	B	B	W
b	B	W	G	R
c	G	R	W	G
d	W	G	R	B

Table 4.5-1

Now look back over this explanation to see that twice we said, "Suppose there is a Hamilton circuit . . . ," and each time we found one. But with a different distribution of colors on the cubes, we might not find two Hamilton circuits of the type we wanted (perhaps not even one). Does that mean that the puzzle has no solution? Not necessarily. What is needed to solve the puzzle is a pair of disjoint sets of four edges labeled a, b, c, d such that in M each set forms at least one of the following.

One circuit of length 4

Two circuits, each of length 2

Two circuits of lengths 3 and 1

Three circuits of lengths 2, 1, and 1

Four circuits, each of length 1

Each listed collection of circuits must contain all four vertices of M, and a circuit of length 1 is simply a loop, a circuit of length 2 is a double edge, and a circuit of length 3 is a triangle. ∎

Exercises 4.5

1. Explain why a separable graph (Section 4.3) cannot have a Hamilton circuit.

2. Show by clear arguments that each of these Platonic solids has a Hamilton circuit. (The same is true for the dodecahedron, as we have seen, and also for the icosahedron.)

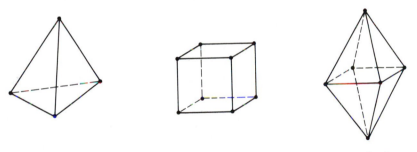

    (a) tetrahedron   **(b)** cube   **(c)** octahedron

3. Explain why a connected bipartite graph with an odd number of vertices cannot have a Hamilton circuit.

4. Observe that the following graph is a modified form of the graph in Figure 4.5-3. Describe a Hamilton circuit, or show that none exists.

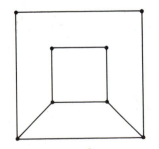

5. Observe that the following graph is a modified form of Hamilton's Icosian graph (Figure 4.5-1). Describe a Hamilton circuit, or show that none exists.

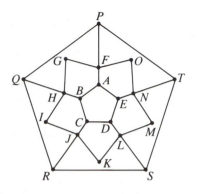

6. Hamilton's Icosian game was accompanied by printed instructions that described some puzzles that start with specified placements of numbers on several consecutive vertices. He also challenged the players to invent further variations.

   (a) With the first five numbers placed at *B, C, P, N, M*, determine a Hamilton circuit in the following graph.

   (b) With the first three numbers on *B, C, D*, determine a Hamilton path that ends at *T*.

   (c) With the first four numbers on *B, C, D, M*, determine a path that uses six additional pieces to become "terminated," meaning that all positions adjacent to the last piece placed are already occupied.

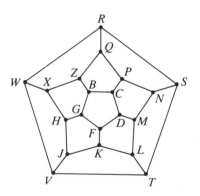

7. For each of the following graphs find a Hamilton path. If a Hamilton circuit exists, find one, or explain why none exists.

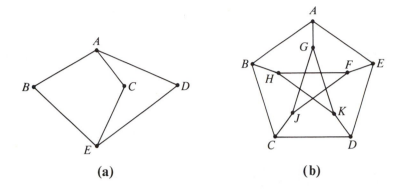

(a)   (b)

8. Let G be a connected bipartite graph of n vertices having a Hamilton circuit. Show that n must be an even number and that each vertex has degree $d \le n/2$.

9. The complete graph K_n with n vertices has a Hamilton circuit for each $n > 2$.

   (a) Prove the assertion.
   (b) If n is odd, then there are $(n - 1)/2$ Hamilton circuits with no edge occurring in two circuits. (You may accept that to be correct.) Demonstrate the result when $n = 5, 7, 9$.

10. A club of 11 students holds its meetings with all members seated at a round table. If any member is absent, the meeting is automatically canceled, and members agree to meet only as many times as it is possible for each member to sit next to a pair of members neither of whom has sat next to him or her at a previous meeting. How many meetings are possible? Show a seating arrangement for each meeting.

11. The table describes a set of cubes for an Instant Insanity puzzle. Showing all of your work and reasoning, solve the puzzle.

Cube \ Face	1	2	3	4	5	6
a	G	W	B	R	G	B
b	G	G	B	W	R	B
c	W	W	W	B	R	G
d	R	B	R	G	R	R

4.6   Uses of Colors in Graphs

We have seen that the early development of graph theory was stimulated by specific problems that were easy to state and sometimes easy to answer correctly, but proofs that the proposed answers were indeed correct were elusive. The most famous problem was stated by Francis Guthrie to his brother Frederick about 1850, when the latter was a student of Augustus De Morgan in London. Francis claimed to have solved the problem, and he outlined his proof orally to Frederick, who told De Morgan about the assertion.

> ***Four-Color Conjecture***   Any map drawn on the surface of a sphere can be colored with only four colors in such a way that any two distinct regions sharing a common boundary segment are assigned different colors.

No map yet drawn had required more than four colors, and the map in Figure 4.6-1 proves that no fewer than four colors can color every map. (This was virtually the only part of Francis Guthrie's supporting argument that was reported in Frederick Guthrie's only publication on this subject, 30 years after Francis left England to live in Capetown.)

De Morgan relayed the problem to his mathematical friends, but little action was taken. In the United States Charles S. Peirce worked on it during the 1860s and thereafter; in 1878, Cayley took up the problem and reported a year later to the Royal Geographical Society concerning the difficulties he encountered in his effort to settle the conjecture. Soon after the problem was solved by a London lawyer, Alfred Kempe (a former student of Cayley), whose proof was published in 1879. Other papers containing alternative proofs were then published. Honors were bestowed on Kempe. Lewis Carroll seized on the idea to devise a map-coloring game. Even the Bishop of London published a proof.

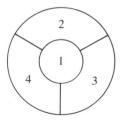

Figure 4.6-1

But in 1890, the bubble burst when a subtle fallacy in Kempe's "proof" was apologetically but unmistakably identified by P. J. Heawood. After careful study of Heawood's paper, Kempe was convinced that his 1879 paper contained a gap in reasoning. Other "proofs" were demolished more easily, and the problem was returned to the status of conjecture. Heawood, however, used Kempe's work to show that five colors will color any planar map. We leave the story at this point to examine its relation to graph theory and return later to describe the contemporary status of the map-coloring question.

Vertex Coloring

Euler showed us how to reduce a map drawn on a sphere to a graph drawn on the plane. With each geographical region we associate a unique vertex so that distinct regions determine distinct vertices. Then two vertices are joined by an edge if and only if the two corresponding geographical regions share a common border of positive length (that is, the common border must contain a segment of points, not just isolated points). The result will be a planar graph, and we want to color all of the vertices, using the smallest possible number of colors such that adjacent vertices have different colors.

EXAMPLES

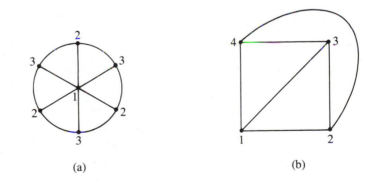

Figure 4.6-2 (a) Three colors suffice; (b) four colors suffice. ∎

Definition 4.6-1

(a) A *vertex coloring* of a graph G is any assignment of colors (one to each vertex of G) such that no two adjacent vertices are assigned the same color.

(b) The *chromatic number* of a graph G is the smallest number of colors needed to color each vertex of G in such a way that adjacent vertices have different colors.

The four-color conjecture then becomes, "The chromatic number of a planar graph is never larger than 4." Probably because of interest in this conjecture there are now many theorems concerning colorings of graphs; in spite of such efforts, however, the problem of finding an efficient algorithm for calculating the chromatic number of an arbitrary graph remains unsolved.

The two graphs of Figure 4.6-2 have chromatic numbers 3 and 4, respectively. The second of these graphs is K_4.

Theorem 4.6-1 K_n has chromatic number n, one more than the degree of each vertex of K_n.

Of course, if any edge is removed from K_n, then the two associated vertices could be colored with the same color, reducing the chromatic number by one.

Theorem 4.6-2 Let G be a graph and let d be the largest degree of any vertex of G. Then the vertices of G can be colored with d or fewer colors if G does not contain a complete subgraph K_{d+1}.

APPLICATION TO SCHEDULING MEETINGS Many organizations work through committees that often have overlapping memberships and usually are expected to meet once a week. When the number of committees is very small or the extent of overlap is small, scheduling is no problem, but, in general, an attempt is made to avoid schedule conflicts even when the likelihood of conflict is large.

Given any list of committees and the members of each, we would like to determine the smallest number of meeting times required for all meetings to be scheduled without conflict. To do so, regard each committee as a distinct vertex in a graph G, with an edge between two vertices if and only if those two committees have a member in common and therefore cannot meet at the same time. The smallest number of meeting times needed to schedule meetings without conflict is simply the chromatic number of G. ∎

Edge Coloring

The use of colors on graphs is largely a matter of convenience for the application at hand. Sometimes it is useful to color edges in such a way that no two edges incident at the same vertex are assigned the same color. The smallest number of colors required is called the *edge-chromatic* number of that graph.

However, colors can be used on edges in many other ways. To be specific, let us interpret a graph G to be a representation of a symmetric binary relation R on the set of vertices of G. The presence of an edge from u to v tells us that $u \, R \, v$. The absence of an edge from v to w tells us that v is not related to w. If we draw each edge of G in red ink and connect each pair of nonrelated vertices in blue ink, the result is the complete graph K_n on the n vertices of G. The red edges belong both to K_n and to G, whereas the blue edges belong to K_n but not to G.

A Germinal Example

The William Lowell Putnam Mathematics Competition is held annually for undergraduates in the United States and Canada. The 1953 examination contained the following problem.

Prove that at a gathering of any six people, some three of them are either mutual acquaintances or complete strangers to one another.

Without our brief discussion of bichromatic edge coloring, we might be puzzled as to how to approach this problem. We first determine what is essential and what is irrelevant in the statement of the problem. It is irrelevant that the six objects are "people," some of whom are "acquainted." It is essential, however, that a symmetric binary relation is defined on a set of six elements. We interpret "mutual acquaintances" to be a related pair and "complete strangers" to be an unrelated pair.

If we represent this binary relation by a graph, we obtain a graph G with six vertices, certain pairs of which are related by being "mutual acquaintances." These pairs are drawn with red ink (solid lines), and the edges between unrelated pairs are drawn with blue ink (dashed lines) to obtain the complete graph K_6 (ignore colors).

Now the Putnam problem asks for a proof that there are three vertices connected in pairs by red edges, or that there are three vertices connected in pairs by blue edges, no matter how the edges of K_6 are colored. An example is shown in Figure 4.6-3.

Thus we are to prove that there must be one monochromatic triangle. In Figure 4.6-3, for example, vertex sets 1, 4, 5 and 1, 3, 4 form red (solid line) triangles, and vertex set 3, 5, 6 forms a blue (dashed line) triangle. In this case, then, the conclusion holds. But we must prove that there must be at least one monochromatic triangle no matter how the edges of K_6 are colored with two colors. To do so, we consider any vertex v. In K_6 each v is of degree 5, so at least three of the edges incident at v are of the same color. We choose any three edges of the same color (say, blue) that leave v and lead to vertices a, b, and c. Edges $\{v, a\}$, $\{v, b\}$, and $\{v, c\}$ are blue. If any of the edges $\{a, b\}$, $\{b, c\}$, $\{a, c\}$ is blue, then that edge with two of the blue

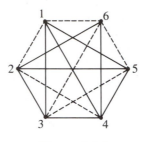

Figure 4.6-3

edges at v form a blue triangle. But if none of the edges $\{a, b\}, \{b, c\}, \{a, c\}$ is blue, all must be red, so (a, b, c) is a red triangle. The proof is complete. ∎

As an exercise, we may show that the edges of K_5 can be colored red and blue in such a way that no monochromatic triangle exists, and so 6 is the smallest number n such that K_n contains at least one triangle with all edges red or at least one triangle with all edges blue. Thus we write

$$R(3, 3; 2) = 6$$

to indicate that 6 is the smallest number n such that whenever the edges (two-vertex subgraphs) of K_n are colored with two colors in any way, K_n contains either a subgraph of the form K_3, all edges of which are red (a monochromatic K_3), or a monochromatic K_3 subgraph, all edges of which are blue.

The symbol $R(k_1, k_2; 2)$ is called a *Ramsey number,* in recognition of the English logician Frank P. Ramsey who proved the existence of such numbers in a 1930 study of formal logic.

Definition 4.6-2   For given positive integers k, c, and k_i ($i = 1, 2, \ldots, c$), let $c \geq k$ and $k_i \geq k$ for each i. The Ramsey number

$$R(k_1, k_2, \ldots, k_c; k)$$

denotes the smallest integer n such that whenever all k-vertex subgraphs of K_n are colored with c colors, there must exist an index i such that K_n contains a complete subgraph G_i with k_i vertices, all of whose k-vertex subgraphs are colored monochromatically with color i.

It will be helpful to restate the significance of each of these symbols. We are thinking of the complete graph K_n having n vertices, where for the moment the value of n is unknown. We choose an integer k and plan to color all k-vertex subgraphs of K_n with c colors, where $c \geq k$. With each color we associate any number k_i such that $k \leq k_i \leq n$. The Ramsey number $R(k_1, k_2, \ldots, k_c; k)$ is the smallest integer n that guarantees that for some index i there is a complete subgraph G_i of K_n having k_i vertices such that all of the k-vertex subgraphs of G_i are colored with the color associated with index i.

Ramsey's study proved that such a number n exists for all values of k and all $c \geq k$ and $k_i \geq k$ for $i = 1, 2, \ldots, c$. But an exact value of $R(k_1, \ldots, k_c; k)$ is known only for a few values of k, c, and k_i. Ramsey theory remains a very active area of current research in graph theory and combinatorics.

Now we return to the four-color conjecture. After he discovered the gap in reasoning of Kempe's proof, Heawood continued to work on the problem for more than 50 years. His inquiries solved the coloring problem for maps drawn on many surfaces other than the sphere. Meanwhile, his work was being supplemented by efforts of other mathematicians, including G. D. Birkhoff, who introduced a concept of reduction that was used by others to show that any map of 35 or fewer countries can be colored with four or fewer colors. In 1950, the German mathematician H. Heesch expressed his opinion that the four-color conjecture could be proved by carefully examining a large finite set of special cases, too large for hand computation but small enough for a high-speed computer.

In 1972, two mathematicians at the University of Illinois, Kenneth Appel and Wolfgang Haken, decided to try a carefully planned and sustained attack on the conjecture, using a computer to examine the 1482 key configurations in meticulous detail. The initial runs were encouraging but indicated the need for modifying the procedures and programs. A second stage of the attack was begun in 1974, and the need for further improvements became clear as the computer itself provided more and more information and guidance. By early 1976, all that remained to be done was to make final analyses both by hand and by computer of a large number of special cases to which all other map configurations can be reduced. The proof was published in 1977 in the *Illinois Journal of Mathematics* in two papers entitled "Every Planar Map Is Four Colorable" (137 pages and more than 1000 hours of computer time).

Is the four-color conjecture now the four-color theorem as Appel and Haken claim? Many mathematicians think so, but others are more cautious. Who can forget that the error in Kempe's proof went undetected for nearly 12 years in spite of scrutiny by some of the world's finest mathematicians? And Kempe's "proof" was not nearly so intricate as Appel and Haken's. Never before has the computer borne so much of the responsibility in the demonstration of a mathematical statement. Is it really an acceptable form of mathematical proof? Only time and careful thought will decide the issue.

Exercises 4.6

1. Determine the chromatic number of each of the following graphs. Show clearly that fewer colors won't do.

(a)

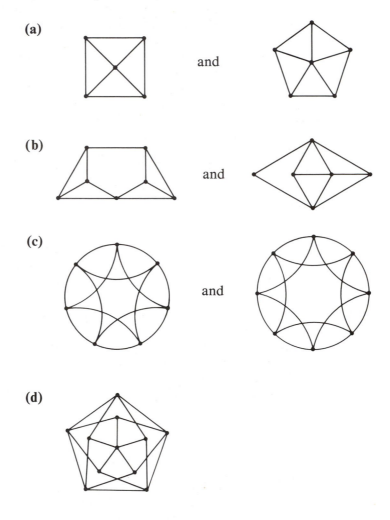

and

(b)   and

(c)   and

(d)

2. Prove carefully that n is the chromatic number of K_n.

3. A graph with $n + 1$ vertices and of the form shown is called a *wheel with n spokes*. Experiment with modest values of n until you feel you have enough evidence to permit you to make a conjecture about the chromatic number of a wheel with n spokes. Then prove your conjecture by careful argument.

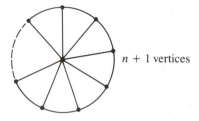

$n + 1$ vertices

4. Prove that a connected graph G has chromatic number 2 if and only if G is bipartite.

5. The Board of Directors of XYZ Corporation has 16 members and 6 committees of 5 persons each as indicated by the following chart.

Committee	Members				
1	A,	D,	G,	K,	N
2	A,	B,	L,	G,	M
3	H,	J,	L,	P,	R
4	C,	E,	F,	M,	Q
5	C,	J,	K,	N,	R
6	D,	E,	H,	P,	Q

How many distinct meeting times are needed to conduct the committee meetings of the board? Show all of your work and reasoning.

6. Determine the edge-chromatic number of each of the following graphs. (No two edges incident at the same vertex may be assigned the same color.)
   (a) The graphs in Exercise 1c
   (b) The graph in Exercise 1d
   (c) The graph in Exercise 6a of Section 4.2

7. Color the edges of K_6 with two colors in such a way that there are exactly two monochromatic triangles.

8. (a) Prove that a graph G can be vertex colored with two colors if and only if G contains no circuit with an odd number of edges.
   (b) Use the results of **a** to conclude that a graph G is bipartite if and only if each circuit of G has an even number of edges.

9. In a round-robin tournament each player plays exactly one match against every other player, and no player may play more than one match a day. How many days are required to complete a round-robin tournament of n players if

(a) $n = 4$,
(b) $n = 5$,
(c) $n = 6$?

10. A manufacturer of chemicals ships its products by railroad tank cars. To reduce the danger that might occur through accidental spills of chemicals, the company specifies that the train must be made up in segments in such a way that

   (a) no two chemicals in the cars of each segment react dangerously with each other,
   (b) two open gondola carloads of sand must precede each segment to separate dangerously reactive chemicals in case of a derailment or other emergency,
   (c) two open gondolas of sand must separate the last segment of chemical cars from the caboose.

   Determine the smallest possible number of gondolas of sand needed to make up a train that carries one tank car of each of the following 12 chemicals. Show your work and reasoning.

   ### Chemicals

   1. Toluene
   2. Acetone
   3. Phosphoric acid
   4. Sulfuric acid
   5. Potassium cyanide
   6. Sodium hydroxide
   7. Dimethyl hydrazine
   8. Dinitrogen tetroxide
   9. Chromic anhydride
   10. Nitrogen
   11. Chlorine
   12. Potassium dichromate

	1	2	3	4	5	6	7	8	9	10	11	12
1	—	S	S	U	S	S	S	U	U	S	U	U
2	S	—	U	U	U	U	U	U	U	S	U	U
3	S	U	—	S	U	U	U	S	S	S	S	S
4	U	U	S	—	U	U	U	S	S	S	U	S
5	S	U	U	U	—	S	S	U	U	S	U	U
6	S	U	U	U	S	—	S	U	U	S	U	S
7	S	U	U	U	S	S	—	U	U	S	U	U
8	U	U	S	S	U	U	U	—	U	S	U	U
9	U	U	S	S	U	U	U	U	—	S	U	S
10	S	S	S	S	S	S	S	S	S	—	S	S
11	U	U	S	U	U	U	U	U	U	S	—	U
12	U	U	S	S	U	S	U	U	S	S	U	—

Reaction Table: S (relatively safe when mixed); U (unsafe when mixed)

4.7   Trees

Section 4.2 described Cayley's work on trees (the word "graph" was not introduced until 1878). Cayley's choice of terminology apparently was generated by the visual forms of the diagrams he drew, because he also used such words as "root," "branch," and "knot."

Definition 4.7-1   A *tree* is a connected graph that contains no circuit (closed trail).

It follows from this definition that a tree has no multiple edges and no loops. Furthermore, a tree is separable if it has more than two vertices (*knots* or *nodes*), and every vertex of degree greater than 1 is a point of separation. Consequently, every tree is planar; because there are no face-separating edges, we have $f = 1$. Thus from Euler's equation it follows that for every tree,

$$1 - e + v = 2,$$
$$v = e + 1.$$

Therefore the structure of trees is unusually simple, with properties that are not generally shared by graphs. In particular, the original terminology that Cayley introduced was borrowed from nature and is still used. A vertex of degree 1 is called a *bud* or *leaf*. Any vertex of a tree can be designated as the *root*; a tree for which a root has been chosen is called a *rooted tree*. Each edge is called a *branch*. Each vertex that is neither a root nor a bud is called a *branch node*. And a set of disjoint trees is called a *forest*. See Figure 4.7-1.

Trees can be defined in a variety of logically equivalent ways, as demonstrated in the following theorem.

Theorem 4.7-1   For any simple graph G the following statements are logically equivalent.

1. G is a tree.
2. G is connected and $e = v - 1$.
3. G has no circuit and $e = v - 1$.
4. There is one and only one path between any two vertices.
5. G is connected, but if any edge is removed the resulting graph is not connected.

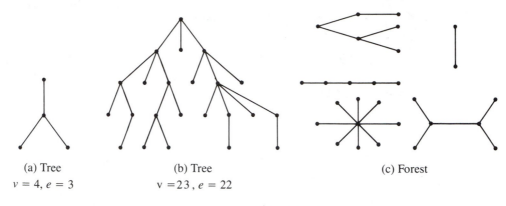

(a) Tree	(b) Tree	(c) Forest
$v = 4, e = 3$	$v = 23, e = 22$	

Figure 4.7-1

To claim that two statements are logically equivalent means that each implies the other. When more than two statements are claimed to be logically equivalent, it is usually convenient (and a great saving of work) to prove a cycle of implications; in this example we shall prove that each statement implies the next, and the last statement implies the first. Then because implication is a transitive relation among statements (see tautology 3 in Theorem 1.7-1), such a cycle shows that each statement implies each of the others.

Proof **1** implies **2**: Let G be a tree. By definition G is connected and has no circuit. As shown earlier, any tree is planar and has no face-separating edges, so $e = v - 1$.

2 implies **3**: Let G be connected with $e = v - 1$. Suppose G has a circuit C; if we remove any edge of C, the resulting graph G_1 is connected, has $e - 1$ edges, and has v vertices. From Theorem 4.2-2 we know that

$$v - 1 \le e - 1.$$

Using the data of statement 2, we obtain

$$v \le e = v - 1,$$

a contradiction: A connected graph with $e = v - 1$ has no circuit.

3 implies **4**: Let G have no circuit, and $e = v - 1$. Suppose vertices U and W exist such that two distinct paths P_1 and P_2 extend from U to W. In proceeding from U toward W along either path, let X denote the first vertex at which the two paths diverge, and let Y denote the first vertex with the property that P_1 and P_2 coincide from Y to W. The edges of P_1

from X to Y followed by the edges of P_2 from Y to X form a circuit, contrary to hypothesis. Hence, at most one path exists between any two vertices U and W of G. To show that one such path exists, we prove that G is connected. Suppose G has k components C_i, $i = 1, 2, \ldots, k$. Then by Theorem 4.2-2 the inequality

$$v_i - 1 \le e_i$$

is valid for each component C_i. Summing over all components, we have

$$v - k = \sum_1^k (v_i - 1) \le \sum_1^k e_i = e.$$

But, by hypothesis, $e = v - 1$, so the value of k must be 1. Hence, G is connected, and a unique path exists from each vertex to any other.

4 implies 5: By hypothesis there is a path in G from each vertex to any other vertex, so G is connected. Consider removing any edge $\{X, Y\}$. If the resulting graph were connected, it would have a path from X to Y. Then G would have a path from X to Y that does not contain edge $\{X, Y\}$, which contradicts the hypothesis that there is a unique path in G from X to Y. Therefore the removal of any edge of G produces a graph that is not connected.

5 implies 1: The last step in this proof is left as an exercise. ∎

Use the examples of trees in Figure 4.7-1 to confirm your understanding of each assertion of this theorem.

EXAMPLES

1. For seven generations the Miller "family tree" was a tree in the sense of graph theory, but several years ago two fourth cousins were married, and now the birth of their child has introduced a circuit in the graph of the descendant relation. So the family tree is now a graph but no longer a tree.

2. Any sequential decision-making process is likely to define a tree. To be specific, consider plans for your spring vacation. See Figure 4.7-2.

3. A tournament (e.g., tennis) is a schedule of contests in which teams or players compete in pairs and one defeat eliminates the loser; it continues until only one player remains undefeated. A tournament is an example of a *binary tree,* which is a tree in which every vertex is of degree 1 or 3. More generally, an *m-ary tree* is a tree in which each vertex is of degree 1 or $m + 1$. See Figure 4.7-3.

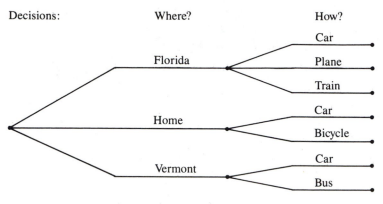

Figure 4.7-2

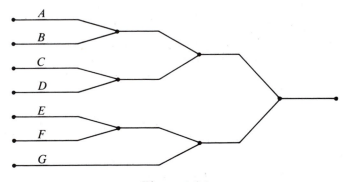

Figure 4.7-3

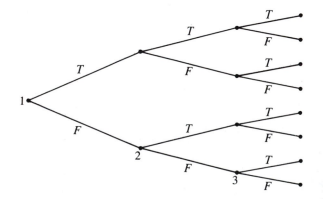

Figure 4.7-4

4. How many different ordered sets of three answers are possible on a true/false quiz of three questions? Each ordered set of three answers is displayed along the branches of the binary tree in Figure 4.7-4. Of course, the number of ordered sets of answers can also be computed by the general Multiplicative Rule of And as $2(2)(2) = 8$, confirming the information in the figure.   ∎

Trees have a special importance in computer science because they are convenient structures for data storage and retrieval. They are also important in graph theory because within each connected graph G there is a tree T that has the same vertex set as G has.

Definition 4.7-2 Let $G = \{V, E\}$ be a graph with vertex set V and edge set E. Any subgraph T of G such that T is a tree and has the form $T = \{V, E'\}$ is called a *spanning tree* of G. (See Definition 4.3-2.)

Observe carefully that a spanning tree T of G has the same vertex set as G. In that sense a spanning tree is analogous to a Hamilton path of G, the distinction being that a Hamilton path is a totally ordered set, whereas a tree is a partially ordered set that is not necessarily totally ordered.

Given a graph G, a spanning tree of G is not hard to construct. Consider the Icosian graph (Figure 4.5-1) as an example. A spanning tree T of G must contain all vertices of G but no circuits. Hence, we repeatedly remove from G one edge at a time, always from a circuit, to make certain that the resulting graph, after each removal, remains connected. In Figure 4.5-1 if we remove the five sides of the outer pentagon and the five sides of the inner pentagon, a connected graph remains but still has one circuit; so if one edge of that circuit is removed, a spanning tree remains. Observe that each edge removal from a circuit eliminates one of the remaining circuits. See Figure 4.7-5.

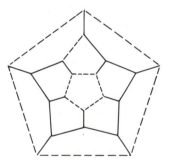

Figure 4.7-5 *A spanning tree.*

The procedure described for this example is perfectly general and can be used to prove the following theorem.

Theorem 4.7-2　Any connected graph has at least one spanning tree.

APPLICATION　Lake Wobegon Airlines has just received approval of their proposal to provide daily commuter service connecting International Falls, East Grand Forks, Bemidji, Moorhead, Duluth, St. Cloud, and Minneapolis. The new airline cannot yet buy enough planes and employ enough personnel to schedule flights between all pairs of cities, so it plans to start with a single route of smallest possible length that visits each of the seven cities exactly once. The route will be flown once each day in each direction. If the one-way route requires more than one plane, the company will acquire the equipment and personnel needed. Use the information provided in Figure 4.7-6 to determine the best route.

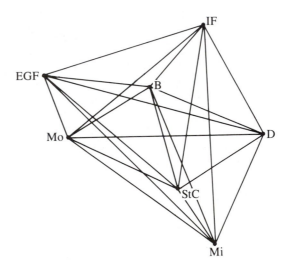

Mileage Chart

	B	D	EGF	IF	Mi	Mo	StC
B	—	132	100	99	179	102	132
D	132	—	234	132	127	205	123
EGF	100	234	—	168	258	72	204
IF	99	132	168	—	238	188	204
Mi	179	127	258	238	—	204	56
Mo	102	205	72	188	204	—	146
StC	132	123	204	204	56	146	—

Figure 4.7-6

The problem posed by the fledging airline is to find a spanning tree of minimum length for the graph K_7 of the seven cities. This general problem has many different settings and variations. Any graph that has a numerical value attached to each edge is called a *weighted graph*. Typically, one is asked to find an optimal path of some type, where "optimal" can be defined in various ways, such as minimal total cost or maximal total revenue, perhaps subject to additional constraints.

There are several good algorithms for constructing a *minimal spanning tree* of a weighted graph — that is, a spanning tree in which the sum of the weights of its edges is as small as the sum of the weights of the edges in any spanning tree. We describe and illustrate a method of R. C. Prim (1957).

Minimal Spanning Tree Algorithm

Step 1. Start with any vertex v_1 of a graph G, and let T denote the subgraph $\{v_1\}$ of G. (For this example start with the vertex labeled StC.)

Step 2. The graph constructed by this process to this point is a tree T. Consider all vertices of T and choose any edge of smallest weight that leads from a vertex of T to a vertex not in T. (The edge from St. Cloud to Minneapolis is the right choice for our example.)

Step 3. Repeat step 2, recognizing that T represents a tree with one more branch (edge) each time step 2 is completed. Continue until each vertex is chosen once. (In this example the vertices adjoined in succession are then D, IF, B, EGF, Mo. See Figure 4.7-7. You should verify that each of these cities is specified in the given order by repeated application of step 2 of Prim's algorithm.)

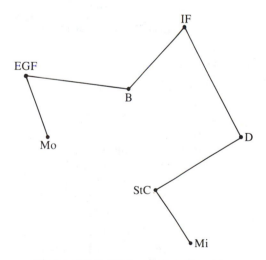

Figure 4.7-7 *Minimal spanning tree.*

In this instance the minimal spanning tree is a path, which is necessarily the case when the underlying graph is complete. In general, however, the minimal spanning tree of a graph can have vertices of degree larger than 2.

Exercises 4.7

1. There are 83 entrants in the community singles tennis tournament.
   (a) How many two-player matches are needed to determine the championship?
   (b) How many "rounds" of matches are needed?
   Show your reasoning clearly for each answer.

2. Explain in careful detail how you can be certain that a graph must be a tree if there is one and only path between each pair of vertices.

3. Explain how you can be certain that when one edge is removed from a circuit in any planar graph, the resulting graph has one and only one fewer region than the graph had before the edge was removed.

4. Using Bemidji as a starting vertex, use Prim's algorithm to construct a minimal spanning tree for Lake Wobegon Airlines and compare it with the route obtained by starting in Minneapolis. Describe what you observe and explain why it is either surprising or not surprising, according to your expectations.

5. In Theorem 4.7-1 show that statement 5 implies statement 1, thus completing the proof of that theorem.

6. Explain how Prim's algorithm can be used to construct a spanning tree for any connected graph whether or not a "mileage" is attached to each edge.

7. In the Lake Wobegon Airlines example let $d(U, V)$ denote the distance between City U and City V. For each choice of U Figure 4.7-6 tabulates $d(U, V)$ for each of the six choices of V. Let $\max_V d(U, V)$ denote the largest of those six distances, the distance from U to its most distant neighbor. As U varies over the list of seven cities, seven such distances are determined. The smallest distance is $\min_U (\max_V d(U, V))$. The city U for which that number is smallest is declared to be the most central of the seven cities, because the distance to its farthest neighbor is least.
   (a) For each choice of U determine $\max_V d(U, V)$ and use that information to determine which of the seven cities is the most central.
   (b) Similarly, for each choice of U calculate $\min_V d(U, V)$, determine

$\max_U(\min_V d(U, V))$, and describe how its value is related to the number $\min_U(\max_V d(U, V))$.

(c) State in geographic terms a title that describes the city U for which $\min_V d(U, V)$ is greatest.

4.8 Search Strategies and Mazes

Search Procedures

One of the world's earliest bull stories comes from ancient Greek mythology. In the city of Knossos on the north coast of Crete lived King Minos whose wife had given birth to the Minotaur, a monster that was half man and half bull. The king commissioned the master builder, Daedalus, to build an underground maze, known as the labyrinth of Knossos, in which to house the Minotaur and the seven youths and seven maidens who were sent from Athens to Crete annually as tribute to the king, eventually to be killed by the Minotaur. A young Athenian hero named Theseus volunteered to be sent to Crete as part of the annual payment. When he arrived, he caught the fancy of Ariadne, the daugher of Minos, who gave Theseus a ball of thread that he could take into the maze. He could then kill the Minotaur and, by rewinding the thread, lead the captives out of the labyrinth to freedom.

It is such a ball of thread that we now seek to spin. Traversing a maze is undeniably a decision process. Every junction of passageways calls for a decision about which path to follow from there. Each passageway leads to a new junction or to a dead end, from which the only option is to retrace one's steps at least to the previous junction. In short, by regarding each junction of passageways as a vertex, each passageway as an edge, and each dead end as a vertex of degree 1, we can represent a maze as a connected graph G. A spanning tree of G visits each vertex exactly once, and therefore it describes a vertex-by-vertex search for the Minotaur (provided that the Minotaur does not change position and that from each vertex every incident edge can be scanned completely). We need to recognize that there are various reasons for conducting a search, and any search procedure must be selected with those reasons in mind. Some procedures visit each vertex of a given graph without passing along each edge. Others pass along each edge of G and thereby automatically visit each vertex. Presently we shall consider a search procedure that traverses each edge twice, once in each direction.

Two principal strategies that can be used in searching a connected graph systematically are called *depth-first* and *breadth-first*. In each strategy some vertex is designated as the initial center of the search and labeled O. Further, we assume that the searcher has local sight only and has no knowledge of what lies beyond the limits of sight.

1. As the name suggests, a breadth-first search of the vertices of G begins by spreading out from O to visit each vertex adjacent to O. The next stage of the search repeats the same procedure with new centers of search, using in turn each vertex first visited on the previous stage of the search. The process continues until each vertex of G has been visited exactly once. Then the vertex search is complete.

2. Similarly, a breadth-first search of the edges of G first examines each edge (not previously traversed) incident with the center of search. The next stage always repeats this procedure, with the center of the search moving in succession to the terminal vertex of each edge examined on the previous stage.

3. In contrast, a depth-first search of the vertices of G moves from the center of search (initially O) to some vertex v adjacent to the center, but which has not been visited previously. Then v immediately becomes the new center of search, and the same process is repeated, thereby constructing a path P. Inevitably, at some point the path cannot be extended farther; if each vertex has been visited, the search has ended; otherwise, the center of search *backtracks* along the path P until, for the first time, it is once again adjacent to some vertex w that has not been visited by the search. Then the entire process starts over from that position and builds a new path through w and perhaps other previously unvisited vertices until the process is forced to backtrack again, and so on, until all vertices have been visited.

4. A depth-first search of the edges of G proceeds from the center of search (initially O) along some incident edge that has not been previously traversed in that direction, and the terminal vertex of that edge immediately becomes the new center of search. Then the process is repeated, moving from the new center of search along an incident edge that has not yet been traversed in that direction. Usually the center of search reaches a vertex from which it must *backtrack* along its path to search edges that it had passed earlier.

Further details of a breadth-first edge search of G are given in the following example, the principal ideas of which were known a century ago as a method for traversing all the corridors of a maze. This method could be used equally well to tour an art gallery in such a way that we are certain to view each object on exhibition. We assume that there is a single lobby through which all visitors enter into and exit from the exhibition areas.

An example is sketched in Figure 4.8-1(a). Any such floor plan can be represented as a graph by letting each room be denoted by a vertex; two vertices are connected by an edge if and only if a doorway connects the two rooms corresponding to those vertices. For example, the gallery of Figure 4.8-1(a) can be represented by the graph in Figure 4.8-1(b).

To have an opportunity to see each painting, we need to enter each room in which art is displayed; that can be ensured by passing through each doorway of those rooms. To avoid aimless wandering, we should avoid passing through any doorway more than once in each direction. Also we must begin and end our tour in the lobby, so it is essential at all times to have reserved a route of return to the lobby. In the depth-first algorithm we will show that these conditions can be met for any connected graph G by a depth-first search of the edges of G that starts and ends at a specified vertex and passes along each edge exactly once in each direction. To keep track of which edges have been traversed and in which direction, we shall apply color-coded labels at the start and end of each edge as it is being traversed: red at the start of each directed passage, and either blue or yellow at the end of that passage, as specified in the algorithm. At the conclusion of the tour each edge will bear two colors at each vertex: red and blue, or red and yellow. As usual, red signifies stop; by placing a red label at the start of each directed edge that we traverse, we record the fact that that edge has been searched in that direction, so we will know not to start along an edge that already bears a red label. Furthermore, yellow signifies a need for caution; we place a yellow label at the end of a traveled edge only when that terminal vertex has not been entered previously by the search, thereby reserving that edge for a later return to O. Thus the searcher will be warned never to start along an edge that bears a yellow label until every other edge at that vertex bears a red label.

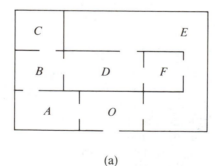 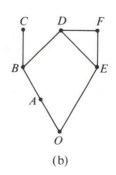

(a)   (b)

Figure 4.8-1

A Depth-First Algorithm for an Edge Search Let G denote a connected graph or multigraph in which a starting vertex is designated by O. Let U and V denote adjacent vertices, and let (U, V) denote the directed edge from U to V. Initially no colored labels have been assigned.

Step 1. Let U be O, and let V be the terminal vertex of any directed edge (U, V). Color (U, V) red at U and yellow at V.

Step 2. Now let U denote the terminal vertex of the directed edge that was most recently traversed. Consider the set $E(U)$ of all edges incident at U that are not colored yellow at U.

a. If every edge in $E(U)$ is colored red at U, go to step 3.

b. Otherwise, some edge (U, V) in $E(U)$ is not colored red at U. Select (U, V) as the next edge of the tour, and color (U, V) red at U. If the tour has not visited V previously, color (U, V) yellow at V. Otherwise, color (U, V) blue at V. Now repeat step 2.

Step 3. The algorithm reaches this step only when condition **a** holds in step 2. Let U be the vertex at which the tour is presently located.

a. If U is not the initial vertex, all but one of the edges incident at U bear two colors at U, red and blue; the exception bears only one color, yellow. Backtrack along that edge from U to the opposite vertex V, coloring that edge red at U and blue at V. Return to step 2.

b. Otherwise, U is the initial vertex O. If some edge is incident at O but is not colored red at O, the tour can be continued; go to step 2. If all edges that are incident at O are colored red at O, the tour has been completed.

EXAMPLE 1 We demonstrate this search algorithm by listing a possible tour of the art gallery in Figure 4.8-1. See Table 4.8-1. First draw the graph; as each directed edge is traversed, label it R at its initial vertex and Y or B at its terminal vertex. ■

A Breadth-First Search of the Vertices of G

Now let's return to the notion of a breadth-first search of the vertices of a graph G, which was the first of the four forms of graph searches listed near the beginning of this section. For comparison we shall again use the graph of Figure 4.8-1(b) to demonstrate such a search. The search begins at O and spreads out from there in stages, reminiscent of the way ripples spread outward over the surface of a pond after a trout jumps to snare a fly.

To keep track of the progress of the search, we again use labels, but instead of colors we now assign natural numbers as labels of vertices to

Center of Search	Next Vertex	Terminal Color	Comment (see below)
O	*A*	*Y*	1
A	*B*	*Y*	
B	*C*	*Y*	1
C	*B*	*B*	2
B	*D*	*Y*	2
D	*E*	*Y*	1
E	*F*	*B*	1
O	*E*	*B*	3
E	*F*	*Y*	
F	*D*	*B*	
D	*F*	*B*	4
F	*E*	*B*	
E	*D*	*B*	
D	*B*	*B*	
B	*A*	*B*	
A	*O*	*B*	
O	Stop		

Table 4.8-1

Comments
1. An alternative choice of edge is available.
2. A tour moves against a yellow warning only when no alternative exists.
3. The tour must backtrack; all other edges at *O* are colored red.
4. *DE* is red and *DB* is yellow at vertex *D*, so *DF* is the only choice.

indicate that each labeled vertex has been visited and to record the stage of the process in which each vertex was visited. Initially the search starts in stage 0 at *O*, and *O* is assigned 0 as a label. In stage 1 the search visits each vertex adjacent to *O* and assigns the label 1 to each. Suppose that the search has just completed stage *k*, and let *S(k)* denote the set of vertices that have *k* as a label. Then in stage *k* + 1 each unlabeled vertex adjacent to some vertex of *S(k)* will receive the label *k* + 1. The process ends when all vertices have been labeled.

EXAMPLE 2 This breadth-first vertex search for the graph of the art gallery can be presented clearly and compactly as in Table 4.8-2. Observe that the label assigned to each vertex is the number of the stage of the search in which that vertex received its label, which also is the smallest number of edges in any path from that vertex back to O.

Stage Number k	Vertices Labeled	Label	Adjacent, Unlabeled
0	O	0	A, E
1	A, E	1	B, D, F
2	B, D, F	2	C
3	C	3	None

Table 4.8-2   ■

We now state the algorithm more formally.

A Breadth-First Algorithm for a Vertex Search Let G be a connected graph with a designated vertex O. The following procedure labels each vertex V of G with the natural number that specifies the smallest number of edges in any path from O to V.

Step 1. Assign the number 0 as the label of the designated vertex O.

Step 2. Let k be the largest number that has been assigned as a label to any vertex of G, and let $S(k)$ denote the set of all vertices labeled k. Let $A(k)$ denote the set of all unlabeled vertices X such that X is adjacent to some vertex in $S(k)$.
a. If $A(k)$ is not empty, assign the label $k + 1$ to each vertex in $A(k)$. Repeat step 2.
b. If $A(k)$ is empty, all vertices have been labeled. Stop.

Finally, we note that a spanning tree of a connected graph G can be constructed by either breadth-first or depth-first searches of the vertices of G, but usually the results will be quite different; the former tree is broad and short, whereas the latter is long and narrow. This remark is intended only to reemphasize that a choice of search strategy should take into account any information available concerning the structure to be searched.

Exercises 4.8

1. Use the labels given in the graphs to apply the breadth-first algorithm for a vertex search of each of the following graphs. Use vertex O as the initial vertex in each case.

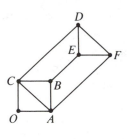

(a)

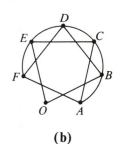

(b)

(c)

(d)

2. Ancient coins from Knossos have this pattern on one side. Though the hypothesis is doubtful, the claim is sometimes made that this represents the labyrinth of Knossos. Draw a graph that represents this maze and use it to state your opinion about that hypothesis.

3. Apply the depth-first algorithm for an edge search of a graph G to each of the graphs given in

(a) Exercise 1a,

(b) Exercise 1c.

4. The famous maze at Hampton Court is shown in the diagram.

    (a) Draw and label a graph that represents that maze, using the letters shown in the maze as labels for corresponding vertices.

    (b) Apply the breadth-first algorithm for a vertex search of that graph.

    (c) Apply the depth-first algorithm for an edge search of that graph.

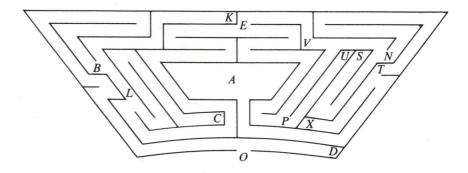

5. Various positions in the following maze are labeled in a somewhat haphazard fashion. Use those labels to represent the maze by a graph *G*. Then describe a path in *G* from *O* to *M* that has the smallest possible number of edges. State clearly how you know that no such path has fewer edges.

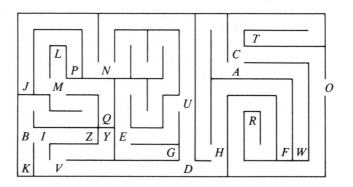

Chapter 5

GRAPHS:
APPLICATIONS AND
ALGORITHMS

This chapter extends our study of graphs and considers several applications of graph theory to problems of current interest. It also gives algorithms for solutions to these problems. Typically, the analysis of such problems involves many calculations, comparisons, and decisions. We are interested not so much in finding a solution as in choosing, out of a large set of solutions, an *optimal* solution—one that is best in some well-defined sense.

5.1   Weighted Graphs

Suppose we want to determine the best way to travel by automobile from Birmingham, Alabama, to Boise, Idaho. "Best" might be defined as "most scenic," "smallest distance," "least elapsed time," "most friends and relatives visited en route," or "minimal mountain driving." Clearly, there are many different routes to choose from, regardless of the criterion for "best," and an unequivocal measure of each route relative to that criterion is needed to choose between different routes. To provide a clearly defined measure of each route from Birmingham to Boise, we utilize the concept of a *weighted graph,* a graph in which a numerical value is assigned to each edge.

> **Definition 5.1-1**   Let G be a graph with n vertices labeled 1, 2, . . . , n. An edge of G from vertex i to vertex j can be denoted by the pair $\{i, j\}$. Suppose w is a function that assigns to each edge $\{i, j\}$ of G a number $w(i, j)$. Then G is called a *weighted graph,* and $w(i, j)$ is called the *weight* of edge $\{i, j\}$.

In various applications we might use other terms for $w(i, j)$, terms that seem more appropriate than "weight" in a particular context. If we refer to an edge by the symbol e_k, then we might write $w(e_k)$ in place of $w(i, j)$. We shall make two simplifying assumptions — one concerning the weight function w, the other concerning G.

1. The value of $w(i, j)$ is a nonnegative integer.

2. The graph G is connected and has no loops.

An illustration of a weighted graph is given in Figure 5.1-1, where, for example, $w(A, C) = 8$ and $w(C, E) = 2$.

EXAMPLE 1   In the discussion of spanning trees in Section 4.7, we considered the complete graph G of airline routes connecting all pairs of seven cities in Minnesota, and we constructed a minimal spanning tree for that graph, a connected set of edges of G that connects all vertices of G and is of smallest total length. G is a weighted graph in which the weight of each edge is the airline distance between the cities connected by that edge. In that example the minimal spanning tree turned out to be a Hamilton path from Minneapolis to Moorhead with stops at five intermediate cities. Other examples, however, show that a spanning tree is not necessarily a path. Section 4.7 gives an algorithm for constructing a minimal spanning tree.   ■

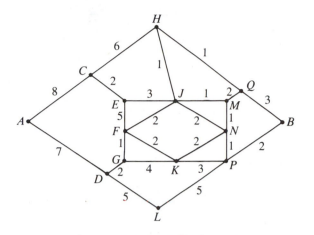

Figure 5.1-1 *A weighted graph.*

EXAMPLE 2 One of our first examples of a graph was introduced in Section 2.2 to model the web of influence between five persons serving together on a committee. We can regard that graph (Figure 2.2-2) as a weighted directed graph in which $w(i, j) = 1$ for each edge (i, j). A more sophisticated model of influence within the committee would be obtained by assigning to each directed edge into vertex j a number $w(i, j)$ such that

$$\sum_{\text{all } i} w(i, j) = 1 \qquad \text{for each } j, \qquad 0 \le w(i, j) \le 1.$$

The number $w(i, j)$ can be interpreted as approximating the proportion of influence on person j that is exerted by person i. ∎

EXAMPLE 3 **A Transportation Problem**

A manufacturer of personal computers has three plants at which computers are produced and five distribution centers from which the computers are available to retailers. Each plant P_i has a given monthly production of p_i units, $i = 1, 2, 3$, and each distribution center D_j has a monthly demand of d_j, $j = 1, 2, \ldots, 5$. The cost per unit of shipping from P_i to D_j is $c(i, j)$ for $i = 1, 2, 3$ and $j = 1, 2, \ldots, 5$. We assume that the combined production is at least as large as the combined monthly demand — that is,

$$\sum_{i=1}^{3} p_i \ge \sum_{j=1}^{5} d_j.$$

This information can be represented by a weighted bipartite graph as illustrated in Figure 5.1-2, where we write $P_i(p_i)$ and $D_j(d_j)$ as labels of the

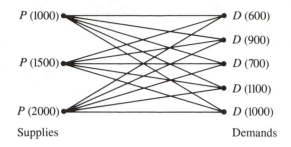

	To					
	Shipping Costs per Unit					
From		D	D	D	D	D
P		1	3	2	3	5
P		2	1	6	3	2
P		1	4	3	4	2

Figure 5.1-2

vertices. The per-unit shipping costs $c(i, j)$ are given in the accompanying table. The problem is to determine a minimum-cost shipping schedule that will supply the five demands. ∎

EXAMPLE 4 A telecommunications system can be modeled as a weighted graph in which each vertex represents a relay station that can receive messages from other stations and send messages to other stations. Each individual subscriber can send and receive messages through the nearest relay station, but for simplicity we include only the relay stations and their connections with other relay stations in describing the system. Each communication link between stations has a finite capacity $c(i, j)$, which is the number of messages that can be transmitted simultaneously along the link between station i and station j. Messages received at station j either must be diverted to the local destination or switched to the link from station j to station k, if there is unused capacity available on the link (j, k) and if station k can send the message onward toward the intended recipient. Otherwise, station j sends an "all circuits busy" signal back to station i. The problem is to understand how to program automatic equipment to route a vast number of calls correctly and efficiently. ∎

These examples were chosen to illustrate applications of graphs requiring some type of optimization. As we consider a few such problems, we shall be particularly interested in understanding associated algorithms—step-by-step instructions for constructing a solution of the type required.

EXAMPLE 5   We can use the setting of Example 1 to pose another problem, which is considerably harder to solve than that of constructing a minimum spanning tree for a weighted graph, as we did in Section 4.7 for the Lake Wobegon Airlines. The Minnesota field representative for a paper company is expected to visit customers in those seven cities on a somewhat regular basis. His home may be in any of those cities. He wants to arrange a tour that starts and ends at his home city and takes him to each of the other six cities exactly once before returning home. Furthermore, he prefers a route that will minimize the total distance traveled. ■

This problem can be stated more simply: *Given a weighted graph, construct a Hamilton circuit of minimum total weight.* This problem has been studied extensively since 1950 and is known as the *traveling salesman problem* (or TSP). Numerous algorithms have been developed that will construct a minimum Hamilton circuit for any weighted graph in which at least one Hamilton circuit exists. However, none of these algorithms is regarded as wholly satisfactory, because none possesses the property of being *computationally efficient.* A precise description of what is meant by computational efficiency is too complex to be included here, so we shall try only to convey the general ideas involved. To apply any computational algorithm to a specific problem, we need input data that describe that problem. Suppose n pieces of input data are required for the execution of an algorithm. Then the "cost" of applying the algorithm can be regarded as a function C of n, where "cost" is defined in terms of the number of arithmetic operations performed, the amount of computer storage space required, computer run time, or similar criteria. If C is a *polynomial* function of n, then the algorithm is regarded as being computationally efficient, and therefore satisfactory. But if C is an *exponential* function of n, the algorithm becomes prohibitively costly to apply as n increases, and the algorithm is said to be computationally inefficient and, therefore (in that sense), unsatisfactory.

A number of problems in graph theory and combinatorics share with TSP the properties of having one or more algorithmic solutions but none that is computationally efficient. Many of these problems have significant economic applications and constitute an active area of contemporary mathematical research. For such problems there is interest also in finding algorithms that produce approximate (rather than exact) computationally efficient solutions. With any approximation method it is useful to have an estimate or upper bound of the magnitude of error for that method. We now state and illustrate two approximation algorithms for TSP. Each algorithm can be used on any weighted graph that has a Hamilton circuit, if the weight function satisfies the condition that for all vertices i, j, and k,

$$w(i, k) + w(k, j) \geq w(i, j).$$

When $w(i, j)$ is interpreted as the distance from vertex i to vertex j, this condition is simply the *triangle inequality* of Euclidean geometry.

Nearest-Neighbor Algorithm

Step 1. Choose any vertex as the starting vertex.

Step 2. Let x denote the vertex most recently chosen. From x proceed along any incident edge of smallest weight to an adjacent vertex that has not been chosen previously.

Step 3. Repeat step 2 until each vertex has been visited exactly once. Return on the edge from the last vertex visited to the starting vertex. Stop.

The existence of the edge referred to in step 3 is guaranteed by the assumption that the graph has a Hamilton circuit, the assumption that the weight function satisfies the triangle inequality, and the method of choosing edges given in step 2. Keep in mind that this algorithm is not guaranteed to construct a *minimum* Hamilton circuit but only to produce a Hamilton circuit that approximates one of minimum length. To illustrate this algorithm, refer to Figure 4.7-6 and choose EGF as a starting vertex. From the mileage table we see that the next choice is Mo, and then B, and so on, as abbreviated in the following table.

EGF		Mo		B		IF		D		StC		Mi		EGF
	72		102		99		132		123		56		258	

The total length of this circuit is 842 miles. In Exercise 5 we will compare this length with the results obtained by starting at other cities.

Nearest-Insertion Algorithm

Step 1. Choose any starting vertex and form a simple circuit C that contains that vertex. (A fictitious loop from that vertex to itself will suffice.)

Step 2. Select a vertex y that is not yet in C; choose y so that the distance from y to the nearest vertex of C is no larger than the distance from any other vertex v not in C to its nearest vertex in C.

Step 3. To choose the position at which to insert y in C, consider all pairs of adjacent vertices u and v in C; select u and v to minimize

$$w(u, y) + w(y, v) - w(u, v),$$

which is the nonnegative increase in the length of the simple circuit C caused by inserting y into C between u and v. (If C consists of a loop from one vertex x to itself, insert y in C at any position other than x.)

Step 4. Repeat steps 2 and 3 until each vertex is in C. Stop.

Although the nearest-insertion algorithm is more complex than the nearest-neighbor algorithm, there is an accompanying reward in that the insertion procedure will always produce a Hamilton circuit whose length is guaranteed not to exceed twice the minimum length of all Hamilton circuits of the given weighted graph.

We illustrate with the same example that we used for the nearest-neighbor algorithm, with C initially being a simple circuit from EGF to Mo to B and return to EGF. The length of C can be determined from Figure 4.7-6 to be $72 + 102 + 100 = 274$. Step 1 is completed.

For step 2 we need to choose the city closest to C but not already in C. By referring to the mileage chart of Figure 4.7-6 and scanning the rows corresponding to the cities in C, we observe that IF is 99 miles from B, whereas other cities not in C are more than 99 miles from all cities of C. Thus we choose IF as the next city to be inserted in C.

In step 3 we decide where to insert IF within C in order to increase the length of C as little as possible. If IF is inserted between EGF and Mo, the increase in length is $168 + 188 - 72 = 284$ miles. If IF is inserted between Mo and B, the increase is $188 + 99 - 102 = 185$ miles. If IF is inserted between B and EGF, the increase is $99 + 168 - 100 = 167$ miles. Hence, IF is inserted between B and EGF, and the new simple circuit C becomes

EGF   Mo   B   IF   EGF

with total length $274 + 167 = 441$ miles. (Before proceeding, make sure that you understand how to obtain the data for these calculations.)

Step 4 instructs us to repeat the calculations and procedures of steps 2 and 3 and augment C by introducing one more city on each cycle until all cities have been inserted into C. We report only the results of those calculations and urge you to verify them by performing the steps of the algorithm. Returning to step 2, we note that D and StC are equally close to the current simple circuit C, and we arbitrarily select D as the next city to be inserted in C. The smallest increase in length occurs if D is inserted between B and IF, and the total length is increased by 163 miles to create the new simple circuit

	EGF	Mo	B	D	IF	EGF

with total length 604 miles. The next cycle of calculations in steps 2 and 3 results in inserting StC between B and D, adding another 123 miles to the expanding simple circuit. The final application of steps 2 and 3 inserts Mi between StC and D, with an increase of 60 miles, to obtain a Hamilton circuit

	EGF	Mo	B	StC	Mi	D	IF	EGF

having a total length of 787 miles. Now step 4 directs us to stop. In Exercise 6 we will compare this length with the results obtained by starting at other cities.

Exercises 5.1

1. Refer to the weighted graph in Figure 5.1-1 and interpret the weights as distances between adjacent vertices. Use the vertex labels to describe a path, as short as possible, from A to B.

2. Refer to the weighted graph shown in Figure 5.1-1 and interpret the weights as flow capacity (in thousands of gallons per hour) of a network of pipes carrying jet fuel to various airports from a refinery located at A. If no other airport is taking fuel from the line, describe the route of a flow, as large as possible, that is available at

   (a) Airport H,
   (b) Airport K,
   (c) Airport C.

3. Give an example of a graph G and a spanning tree T of G such that T is not a path of G.

4. Work out a shipping schedule that meets all of the conditions of Example 3. Attempt to obtain a schedule of lowest total cost, and compute the cost of your schedule.

5. Refer to the graph in Figure 4.7-6 with the accompanying mileage table. Determine the length of the Hamilton circuit constructed by the nearest-neighbor algorithm starting at

   (a) Bemidji,
   (b) St. Cloud.
   (c) Compare those lengths with the corresponding length obtained in the discussion following the statement of the nearest-neighbor algorithm and comment on what you observe.

6. Follow the instructions of Exercise 5, but use the nearest-insertion algorithm instead of the nearest-neighbor algorithm. Also compare the lengths obtained with those obtained in Exercise 5.

7. Refer to the weighted graph in Figure 5.1-1.
   (a) Write a Hamilton circuit for that graph.
   (b) Apply the nearest-neighbor algorithm, starting at vertex A, and listing each vertex along your path until you can proceed no farther.
   (c) State why the failure of the algorithm to produce a Hamilton circuit in this example is not inconsistent with the text presentation.

5.2 Matching

PROBLEM The instructor of a course in composition and literature prepared a list of 32 book titles and asked each of the 23 class members to select one or more titles that he or she would be interested in reviewing for the required term paper. The instructor promised to assign to each student one of the titles that the student had selected, but no two students were to be assigned the same book. After receiving a selection list from each class member, the instructor was puzzled to find that the promise could not be kept even though the combined class selections included all 32 titles. So the instructor added 14 more titles and repeated the individual selection procedure, this time with success. Explain precisely why the second trial succeeded and why the first one failed. ■

A superficial response can be provided in terms of probability: With more books to choose from, it is more likely that the class selection will permit assignment of 23 different titles, each selected by the person to whom it is assigned. But that avoids the issue, because a successful outcome of the process is possible, even with as few as 23 books, so the question really concerns the nature of the set of 23 selection lists submitted by the students. For example, if two students with very particular reading tastes each selected the same book and neither provided an alternative, the assignment process would have failed even if the instructor's list had included 1000 titles. For the assignment to succeed it is *necessary* that the combined selections of each pair of students include at least two titles. By the same reasoning, the combined selections of every three students must include at least three titles if a different book is to be assigned to each student. In general, an assignment will be impossible if for any set of k class members the combined selections include fewer than k titles.

Having discovered a necessary condition for the assignment procedure to succeed, we wonder if the same condition is *sufficient* to guarantee that the process will work. We assume that the combined selection list for each subset of k students out of a class of n students contains at least k different titles, for each positive integer k from 1 to n. It is natural to think of a proof by induction, which is listed as an exercise at the end of this section.

To phrase this problem generally in mathematical form, we are given two finite sets A and Z (students and books in the given setting) and a binary relation R from A to Z (a selection by each member of A of one or more members of Z). The problem is to define a function f with domain A and range in Z such that each of the following conditions is true.

1. f is one-to-one.
2. If $z = f(a)$, then $a \, R \, z$.

Such a function defines an assignment of one book to each student so that (1) no two students are assigned the same book, and (2) the book assigned to each student is from the selection list of that student.

In this form, the problem is applicable in a wide variety of settings. The answer is that a function with the desired properties exists if and only if for each k-element subset S of A there exist at least k distinct elements of Z that are related to one or more elements of S. Indeed, this theorem was first stated and proved in 1935 by Philip Hall in a paper entitled "On Representation of Subsets." In set-theoretic terms a selection of one element from each of a finite collection of sets is called a *system of distinct representatives* of that collection if and only if no two sets in the collection have the same element selected as a representative. Thus Hall's result can be phrased as follows.

Theorem 5.2-1 Let $C = \{S_i\}$ be a collection of n nonempty subsets of a given set Z. Then C has a system of distinct representatives $\{x_i\}$ such that $x_i \in S_i$, and $x_i \neq x_j$ whenever $S_i \neq S_j$ if and only if for every integer k between 1 and n (inclusive) the union of every family of k of those subsets contains at least k elements.

The statement of this problem (using a binary relation R from A into Z to extract a one-to-one function from A into Z) reminds us of the representation of a binary relation by a bipartite directed graph with $A \cup Z$ as the vertex set and with each edge directed from a vertex in A to a related vertex in Z. Conversely, any connected, bipartite directed graph G (that is not a multigraph and has no loops) represents a binary relation from the set A consisting of all initial vertices of the directed edges of G onto the set Z of all terminal vertices of those edges. Hall's theorem can be stated in this context as follows:

> Let G be a bipartite directed graph with a finite initial set A and a terminal set Z (where G has no loops or multiple edges). For each subset X of A let $R(X)$ denote the set of all y in Z such that (x, y) is a directed edge of G for some x in X. There exists a set of edges of G that describes a one-to-one function f with domain A and range Z if and only if for each subset X of A the number of elements in $R(X)$ is at least as large as the number of elements in X.

This form of Hall's theorem reveals why it has become known as "the marriage theorem." Suppose each unmarried woman in a village were to list the name of each eligible man in the community whom she regards as a suitable husband. Then Hall's theorem states a necessary and sufficient condition for the matchmaker to have the possibility of complete success.

Recall that a function f with domain A and range Z assigns to each x in A a uniquely determined element $f(x)$ in Z. Also f is one-to-one if and only if $f(x) = f(a)$ only when $x = a$. When A and Z, respectively, are the initial and terminal vertex sets of a bipartite directed graph, such a function is called a *matching*. The notion of a matching can be stated generally for any graph.

Definition 5.2-1 Let G be a graph (without loops) with E as its set of edges.

(a) A *matching* in G is a subset M of E such that no vertex of G is common to two edges in M. Each edge in M *matches* the two vertices of that edge.

(b) A matching M of G is *perfect* if and only if each vertex of G belongs to an edge of M.

(c) A matching M of G is *maximal* if and only if no other matching of G has more edges than M does.

(d) In a bipartite graph with vertex sets A and Z, a matching from A to Z is *complete* if and only if each vertex of A is matched with a vertex in Z.

For example, the matching whose existence is claimed in the marriage theorem is a complete matching. Examples of matching also are shown in Figures 5.2-1 and 5.2-2, with the edges of M drawn heavier than the edges of G that are not in M. Although the definitions are applicable to any

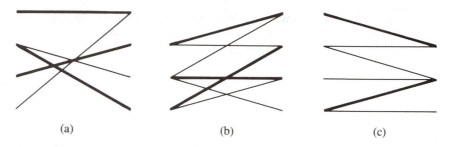

Figure 5.2-1 *Bipartite graphs. (a) Maximal; (b) complete; (c) neither.*

graph without loops, we restrict our attention to the somewhat simpler case of bipartite graphs.

APPLICATION   A school personnel manager plans to employ a bookkeeper, 2 secretaries, 4 bus drivers, 2 custodians, and 3 cooks. There are 14 applicants, several of whom seem to be qualified for more than one of the positions open. The manager might find it useful to sketch a bipartite graph that relates each candidate to each job that he or she is qualified to do, and then to construct a maximal matching for that graph.   ■

This application is a form of the *personnel assignment problem* that was an active area of research in the early 1950s. The following theorem provides a sufficient (but not necessary) condition for a complete matching in a bipartite graph, a condition that often is easy to check.

Theorem 5.2-2   Let G be a bipartite graph with vertex sets A and Z. There exists a complete matching from A to Z if there is a positive integer k such that each vertex of A has degree *at least* k and each vertex of Z has degree *at most* k.

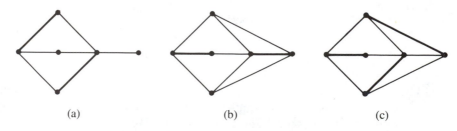

Figure 5.2-2 *(a) Maximal; (b) neither; (c) perfect.*

Proof Let X be any subset of A, and let X contain m vertices. Then at least km edges lead from X to the vertices of $R(X)$ in Z. Each vertex in $R(X)$ has k or fewer incident edges, so $R(X)$ has m or more vertices. Application of the marriage theorem completes the proof. ∎

In particular, this theorem tells the school personnel manager that if each applicant is qualified for at least k positions but no more than k persons apply for any one of the 12 openings, then each position can be filled with a qualified person from the current list of applicants. The value of k can be any positive integer.

From the graph in Figure 5.2-1(b) and from the definitions that precede that figure, we note that in a bipartite graph with vertex sets A and Z we speak of a matching *from A to Z*. The distinction between the two vertex sets is emphasized further in the definition of a complete matching, which requires that each vertex of A, but not necessarily each vertex of Z, be matched. A maximal (but not complete) matching therefore defines a one-to-one function from A into Z with domain A_1, a proper subset of A. From Figure 5.2-1(a) we see that a maximal matching is not necessarily complete, but a complete matching is always maximal.

Although Hall's theorem provides a test for when a complete matching exists, it is usually an inefficient test to apply because it requires that an inequality be verified for every subset of A. Thus it is appropriate to seek an effective test for determining whether a given matching is maximal. We would like also to find a method of extending any matching that is not maximal. Theorem 5.2-3 describes a test for maximality of a given matching M in a bipartite graph. We defer a proof of that test and a method of extending a nonmaximal matching until Section 5.3.

Exercise 6 of Section 4.1 presented a map of a village and asked where as few mailboxes as possible might be located so that each residence had a mailbox at an adjacent street corner. In the language of graph theory the problem is to select a subset K of the vertex set of a graph G so that each edge of G contains at least one element of K. Any such set K is called a *vertex cover* of the edges of G.

APPLICATION   In some art museums paintings and sculptures are displayed along corridors and within rooms. For security reasons each display area is to be within sight of a guard at all times. Hence, a guard is stationed in each room. If a graph G is drawn to represent the museum corridors, a minimal vertex cover K is a smallest set of guard stations needed to provide security along the corridors. ∎

Now let M be any matching of a graph G, and let K be any vertex cover of G. Each vertex in K is incident with at most one edge of M because no two edges of M share a vertex. Hence, the number $e(M)$ of edges in M does not exceed the number $v(K)$ of vertices in K:

$$e(M) \leq v(K).$$

Observe that this is another inequality of the type discussed in Section 2.5. In any graph G we have two types of structures — vertex matchings and vertex covers — with a numerical function defined for each type, such that no value of the first function ever exceeds any value of the second function. Thus for every cover K,

$$\max_{\text{all } M} e(M) \leq v(K),$$

and therefore

$$\max_{\text{all } M} e(M) \leq \min_{\text{all } K} v(K).$$

Seeing this inequality, we are led inevitably to ask, Is it necessarily the case that equality holds for some maximal vertex matching and some minimal vertex cover? Somewhat surprisingly, the answer for connected graphs in general is no. However, for a connected bipartite graph the equality always holds, a result that was established by Dėnes König, in 1931, by constructing a vertex cover K having exactly one vertex for each edge of a maximal matching M.

> **Theorem 5.2-3** Let M be matching from A to B in a bipartite graph G. Then M is maximal if and only if there exists a vertex cover K of the edges of G such that
>
> $$e(M) = v(K).$$

Exercises 5.2

1. State more precisely what is meant by the last sentence in the paragraph of this section that describes Hall's theorem as the marriage theorem.

2. For each part of this exercise let C denote the collection of sets listed in that part. Find a system of distinct representatives for the sets of C, or else show clearly that no such system exists.

(a) $\{a, b, e\}, \{a, b, c\}, \{a, b, d\}, \{a, b, c, e\}, \{a, c, d, e\}.$

(b) $\{c, d\}, \{b, d\}, \{a, d\}, \{b, c\}.$

(c) $\{b, d\}, \{a, d\}, \{a, b, d\}, \{b, d\}.$

(d) $\{a, b, c, d\}, \{c\}, \{d\}, \{c, d\}.$

(e) $\{a, f, h\}, \{e, f, h\}, \{a, e, g\}, \{b, c, d\}, \{c, d, g\}, \{a, e, g, h\}, \{a, e, f, g\}, \{b, d, f\}, \{a, f, h\}.$

3. The executive committee of a foundation consists of five persons and is organized into five committees, each committee having responsibility for planning programs in one area of interest of the foundation, as listed below.

   1. Amar, Berg, Cox

   2. Amar, Davis, Elia

   3. Berg, Cox, Davis

   4. Amar, Cox, Davis

   5. Cox, Davis, Elia

   For budget control each committee reports to a designated overseer from the executive committee, who is not a member of that committee. No person is to be overseer for more than one committee. If the committee assignments are as shown, designate an overseer for each committee, or show that such designations are not possible.

4. When a new Little League baseball team first met with the prospective coach, only nine prospective players attended. The coach asked each player to list two positions that he or she would like to play, with the results shown below. Help the manager by listing all lineups that assign each player to a position that she or he would like to play, or show that none exists.

Position	A	B	C	D	E	F	G	H	J
Pitcher	x	x	x						
Catcher				x	x				
First							x	x	
Second				x					x
Third					x			x	
Shortstop									x
Left			x			x			
Center	x							x	
Right		x				x			

5. At a retirement center 24 persons who enjoy playing cards form a club that meets each Tuesday evening. They play bridge in six groups of four on one Tuesday and poker in four groups of six on the next Tuesday, alternating in this fashion. The groupings remain unchanged for a 12-week interval and then new assignments to groups are made by drawing names. Within any 12-week interval there is to be a committee to facilitate the exchange of information among the members each Tuesday, even though the groups meet in different homes. Is it possible to select a committee by choosing one member in each bridge group so that each poker group also is represented on the committee? Explain fully.

6. As a generalization of the previous exercise, prove that, for any positive integers m and n with $m \le n$, if a set S of mn distinct objects is partitioned into m subsets of size n and then partitioned into n subsets of size m, it is always possible to choose n objects, one from each set of the second partition, so that at least one of those n objects appears in each of the m subsets of the first partition.

7. Explain in detail how the statement of Hall's theorem can be transformed, as claimed in the text, into the given statement about existence of a matching in a bipartite graph.

8. For each of the graphs determine by inspection a vertex cover that you think might be minimal.

(a)

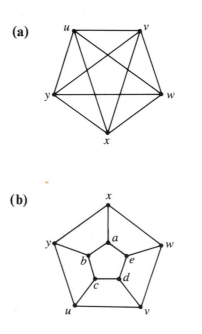

(b)

(c)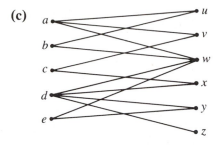

9. Find by inspection in the graphs specified a matching of the vertices that might be maximal.

   (a) Exercise 8**a**,
   (b) Exercise 8**b**,
   (c) Exercise 8**c**.

10. As suggested in the text, use induction to complete the proof of Hall's theorem.

5.3 Constructing a Maximal Matching

This section describes and illustrates an algorithm that either will extend a given matching M' in a bipartite graph G or otherwise will identify a vertex cover K' of the edges of G such that

$$e(M') = v(K').$$

Since the inequality

$$e(M) \leq v(K)$$

holds for every matching M and every vertex cover K, the existence of a vertex cover K' for which equality holds shows that M' is a maximum matching and that K' is a minimum cover. Conversely, if M' is a maximum matching, it cannot be extended; thus the algorithm assures the existence of a vertex cover K' such that $e(M') = v(K')$. Therefore, König's theorem for bipartite graphs will be established by any algorithm that

1. extends a matching that is not already maximum, and
2. determines, for each maximum matching of exactly k pairs of vertices, a set of k vertices that is a vertex cover of the edges of G.

Definition 5.3-1 Let G be a bipartite graph with vertex sets A and Z.

(a) A *chain* is a finite sequence of pairs of vertices of the form

$$\{v_1, v_2\}, \{v_2, v_3\}, \{v_3, v_4\}, \ldots, \{v_k, v_{k+1}\}$$

such that $\{v_i, v_{i+1}\}$ is an edge of G for all i.

(b) The *length* of this chain is k, the number of edges in the chain. A chain of odd length is called an *odd* chain. A chain of even length is called an *even* chain.

Note that in any chain of a bipartite graph, $v_1, v_3, v_5, \ldots$ all belong to one vertex set, whereas $v_2, v_4, v_6, \ldots$ belong to the opposite vertex set. The first vertex of the chain can be in either set. The chain is even if and only if the first and last vertices are in the same set.

Definition 5.3-2 Let M denote a matching of a *bipartite* graph G. A chain in G is *alternating* with respect to M if and only if for every pair of successive edges in the chain exactly one is in M; that is, whenever an edge of the chain is an edge of M, the next edge of the chain is not in M. An alternating chain is an *augmenting chain* with respect to M if and only if the chain is of odd length, neither the first nor the last edge of the chain is in M, and neither the first nor the last vertex of the chain is incident with an edge of M.

Figure 5.3-1(a) illustrates these concepts. The edges of M are shown in solid lines, and other edges of G are shown in dashed lines. The chain $\{b, v\}, \{v, c\}, \{c, x\}$ is an augmenting chain relative to M that contains one edge $\{v, c\}$ of M. Suppose we define a new matching M' to consist of edges $\{b, v\}$ and $\{c, x\}$ together with all edges of M except $\{v, c\}$, as in Figure 5.3-1(b). Then M' has one edge more than M had.

The preceding example is simple, but it clearly reveals the use of an augmenting chain in extending a matching M to obtain a matching M' having one more edge than M has. By definition, an augmenting chain is of odd length and alternating with one fewer edge in M than it has edges not in M. So if we reverse the relation to M of each edge in an augmenting chain, we remove k edges from M and replace them by the $k + 1$ edges of the chain that are not in M, thus obtaining a new and larger matching M'. In this example M' is maximal because each vertex in Z is incident to an edge of M'. A vertex cover $K' = \{x, y, v, w\}$ of the edges of G can be obtained by inspection. In general, however, repeated application of the algorithm eventually will fail to find an augmenting chain, and then the process will define a minimal vertex cover, as desired.

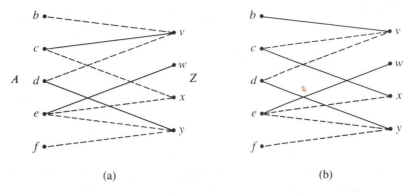

Figure 5.3-1 (a) **M** (solid lines); (b) **M′** (solid lines).

The basic idea of the algorithm is to construct in G all possible chains that are alternating with respect to an initial matching M. Each such chain starts at a vertex in A that is not incident with any edge of M, proceeds to Z along an edge that *is not* in M, and returns to A along an edge that *is* in M, and continues in this way as long as possible without visiting any vertex more than once. Thus, each chain is alternating and must eventually terminate. If some chain terminates in Z, it has an odd number of edges and therefore is an augmenting chain that can be used to construct a new and larger matching in G. Otherwise all chains terminate in A, which signals that M is maximal; moreover, a minimal vertex cover K' of all edges of G can be obtained by forming the union of the set of all vertices of A that *have not* been visited by any chain and the set of all vertices of Z that *have* been visited by some chain. To keep track of all possible alternating chains simultaneously, the algorithm labels each vertex visited, using 0 as a label for each starting vertex, and labeling each subsequent vertex with the symbol for the previous vertex visited by that chain.

A Labeling Algorithm for Maximal Matching (The Hungarian Method)   Let G be a connected bipartite graph with vertex sets A and Z, and let M be a matching in G.

Step 1. Assign the label (0) to each vertex in A that is not incident with any edge of M.

Step 2. Let L_A denote the set of all vertices of A that were labeled in the preceding step. For each a in L_A assign the label (a) to each unlabeled vertex z in Z for which $\{a, z\}$ is an edge of G that is *not* in M. (Repeat for each vertex a in L_A.)

Step 3. Let L_Z denote the set of all vertices of Z that were labeled in the preceding step.

a. If L_z is empty, each alternating chain of M ends in A and has even length; thus no augmenting chain exists, and M is maximal. Go to step 5.

b. If there is a vertex z in L_z that is not incident with any edge of M, then z is the terminal vertex of an augmenting chain. Go to step 4.

c. Otherwise, for each vertex v in L_z there is one and only one vertex b in A such that $\{b, v\}$ is an edge of M. Moreover, b is not labeled. Assign the label (v) to b. (Repeat for each v in L_z.) Return to step 2.

Step 4. The algorithm reaches this step when and only when an augmenting chain C has been found in **b** of step 3. Starting at the terminal vertex of C, trace C back to its origin in A. Define a new matching M' in G as follows:

a. Include only those edges of C that are not in M.

b. Include all edges of M that are not in C.

M' contains one more edge than M. Return to step 1 and use M' in place of M.

Step 5. The algorithm reaches this step only from **a** of step 3 and only when no augmenting chain exists. Hence, the current matching is maximal. Also the union of the set of all unlabeled vertices of A and the set of all labeled vertices of Z forms a minimal vertex cover K' of the edges of G.

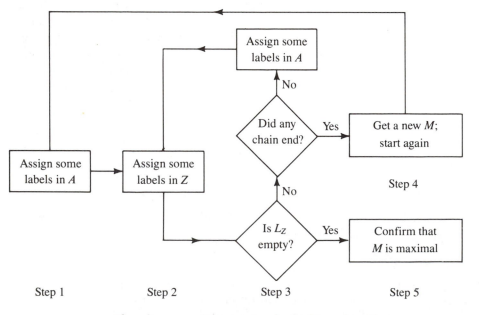

Flow diagram of the maximal matching algorithm.

EXAMPLE   Before we justify the claim made in step 5 of the algorithm, it will be instructive to apply the algorithm to the graph in Figure 5.3-1(a) to obtain the matching in Figure 5.3-1(b) and the vertex cover

$$K' = \{c, e, v, y\},$$

having the same number of vertices as that matching has edges.

When step 1 is applied to G, vertices b and f are labeled (0). Step 2 assigns (b) as a label for v and (f) as a label for y, as shown in Figure 5.3-2(a). Step 3 assigns (v) as a label for c, (y) as a label for d, and returns the process to step 2. Step 2 assigns (c) as a label for x; no other vertex in Z is eligible to be labeled, so the algorithm moves again to step 3. Because no edge of M is incident at x, the labeling process stops in Z, so the algorithm directs us to step 4 with the labels as shown in Figure 5.3-2(b). The sequence of labels identify an augmenting chain $\{x, c\}, \{c, v\}, \{v, b\}$ that can be used to define a larger matching M', as shown in Figure 5.3-2(c).

At this point, step 4 of the algorithm directs us back to step 1 with all labels removed and with the new matching M', as shown in Figure 5.3-2(c). In step 1 vertex f again is assigned (0) as a label; y is labeled (f) in step 2; and d is labeled (y) in step 3. Then steps 2 and 3 are repeated; v is labeled (d), and b is labeled (v). From that point the labeling process cannot continue, and the final label was assigned to vertex a in A. Step 3 directs us to step 5, where we obtain the vertex cover

$$K' = \{c, e\} \cup \{v, y\}.$$

Because the number of edges in M' and the number of vertices in K' are equal, we know that K' is minimal and M' is maximal (Figure 5.3-3). Indeed, the fact that M' is maximal is obvious from Figure 5.3-2(c) because each vertex in Z is matched with one in A.

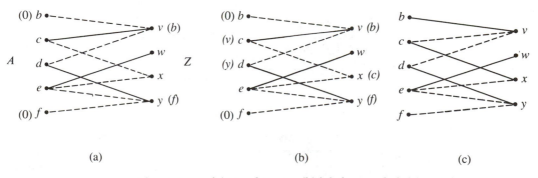

(a)   (b)   (c)

Figure 5.3-2 *(a) Matching* **M**; *(b) labeling ended; (c) matching* **M'**.

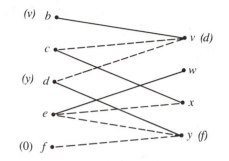

Figure 5.3-3 K' *minimal cover;* M' *maximal matching.*

Now let's return to step 5 of the algorithm to show that the set K' of vertices does cover each edge of G and that $e(M') = v(K')$. From steps 1 and 3 it follows that if a is a labeled vertex in A, then for any vertex z in Z such that $\{a, z\}$ is an edge of G, z must also be labeled. Because G is connected and bipartite, it follows that each unlabeled vertex in Z must be connected by an edge of G with some unlabeled vertex in A. Hence, the set of unlabeled vertices in A covers all the edges that lead to unlabeled vertices in Z. It is evident that the set of all labeled vertices of Z covers all the edges of G that lead into those vertices. Hence K' is a vertex cover of the edges of G. Finally, we need to show that K' has no more vertices than M' has edges. With each edge $\{a, z\}$ in M' we associate one of the two terminal vertices according to this rule:

If z is labeled, we associate z with $\{a, z\}$;
If z is not labeled, we associate a with $\{a, z\}$.

This rule defines a one-to-one function f from the edges of M' onto K' (with the edges of M' as domain), so

$$e(M') = v(K').$$

Finally, observe that the maximal matching algorithm can be used to construct a minimal vertex cover of all edges of a bipartite graph and, thereby, to decide whether a given vertex cover is minimal.   ■

Exercises 5.3

1. Each of the diagrams describes a bipartite graph G with vertex sets A and Z together with a matching M from A to Z. Apply the labeling algorithm to extend M to a maximal matching. Show each labeling sequence that identifies either an augmenting chain or a minimal vertex cover.

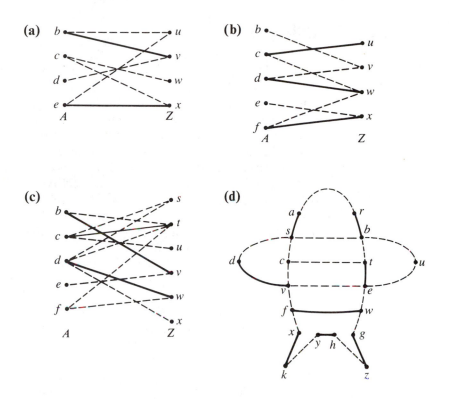

2. At the end of Section 5.3 the assertion was made that when the matching algorithm is applied to a matching M' but comes to a halt without finding an augmenting chain, then M' is maximal because the number of edges in M' coincides with the number of vertices in the vertex cover K' consisting of the unlabeled vertices of A and the labeled vertices of Z. Also the claim is made that these two numbers are equal because of a specified correspondence from the edges of M' to the vertices of K'. Show in full detail that the correspondence is one-to-one and onto K'.

3. Determine a minimal vertex cover and a corresponding maximal matching for each of the following bipartite graphs.

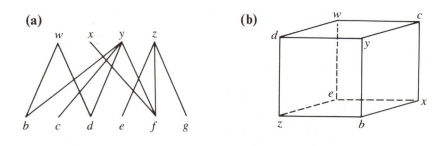

(a)

(b)

5.4   Network Flow

This section considers a type of weighted graph, called a *transport network,* that can be used to model a wide variety of problems of special interest to economists and engineers. Example 3 of Section 5.1 was a specific example of the transportation problem concerned with finding a shipping schedule that will deliver computers from three factories to five outlets in such a way that the demand at each outlet is satisfied, the production capacity of each factory is not exceeded, and the total shipping cost is minimized. The weight attached to each edge of the graph is the cost of shipping one unit along that edge. Regardless of cost, however, the resulting graph depicts a transportation system within which material flows from manufacturing plants (sources) to distribution centers (destinations), subject to constraints imposed by supply and demand. Many other interpretations can be given to such a network, and many other questions can be raised. For example, let P_1, P_2, P_3 denote natural gas wells and D_1 through D_5 denote industrial consumers of natural gas. Each edge $\{P_i, D_j\}$ can be regarded as a pipeline having a specified capacity $c(i, j)$ given by the table in Figure 5.1-2. Moreover, the network might be more complicated than the complete bipartite graph of Figure 5.1-2, perhaps as sketched in Figure 5.4-1.

Along each edge of the network of Figure 5.4-1 there is an expected direction of flow, except along edge $\{M_2, M_3\}$. We shall assign a direction to each edge, and thus obtain a directed, weighted graph. At each P_i every incident edge is directed *away from* P_i. At each D_j every incident edge is directed *toward* D_j. Along any other edge, such as $\{M_2, M_3\}$ in this example, either direction of flow may be assigned.

A description of network analysis can be simplified if we assume each network has a *single* source vertex S and a *single* destination vertex D. But

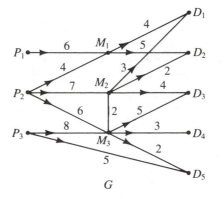

Figure 5.4-1

in Figure 5.4-1, G has multiple sources and multiple destinations, so we modify G by adjoining a single source vertex S with an edge directed from S to each of the sources P_i of G, and by adjoining a single destination vertex D with an edge directed to D from each destination D_j of G. The vertices between S and D and the edges connecting those vertices make an exact copy of G. We shall denote this new directed graph by G_1.

Next, G_1 is made into a weighted graph by adopting without change all weights assigned to the edges of G and by assigning weights to the new edges (S, P_i) and (D_j, D) as follows so that any flow in the graph G can be duplicated in G_1.

1. For each P_i let $c(S, P_i)$ be the sum of the capacities $c(P_i, X)$ for all vertices X adjacent to P_i in G.

2. For each D_j let $c(D_j, D)$ be the sum of the capacities $c(Y, D_j)$ for all vertices Y adjacent to D_j in G.

When this construction is applied to the graph G of Figure 5.4-1, the result is the graph G_1 shown in Figure 5.4-2.

Definition 5.4-1 Let G denote a directed, weighted graph (connected and without loops) in which the weight assigned to the directed edge from vertex i to vertex j is a positive integer, denoted $c(i, j)$ and called the *capacity* of that edge. G is called a *transport network* if both of the following conditions are satisfied.

1. There is one and only one vertex S (called the *source*) such that no edge of G is directed *into S*.

2. There is one and only one vertex D (called the *destination*) such that no edge of G is directed *away from D*.

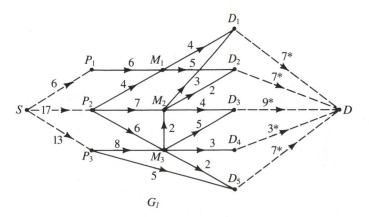

G_1

Figure 5.4-2

Given a transport network G, a question that we frequently want to answer is, What is the maximum amount of material that can flow per unit of time from S to D? Indeed it was precisely this question that attracted the attention of a few mathematicians in the decade following the end of World War II and led in 1956 to one of the early successes of graph theory within the branch of mathematics called *operations research* or, more generally, *optimization*.

Definition 5.4-2 Let G be a transport network with source S, destination D, and intermediate vertices V_k for $k = 1, 2, \ldots, m$. For simplicity we denote the directed edge from V_i to V_j by (i, j).

1. A *flow f* in G is any integer-valued function that assigns to each directed edge (i, j) of G an integer $f(i, j)$ (perhaps negative).
2. A flow in G is *feasible* if and only if two conditions hold:

   a. $|f(i, j)| \leq c(i, j)$ for every edge of G—that is, the number of units of material flowing along any edge must not exceed the capacity of that edge.
   b. Excluding S and D, the total flow into each vertex must equal the total flow out of that vertex — that is,

$$\sum_k f(k, i) = \sum_j f(i, j)$$

   for each vertex i that is neither S nor D.
3. The *value* $v(f)$ of a flow f is the number of units of material per unit of time reaching D—that is,

$$v(f) = \sum_V f(V, D)$$

   for all vertices V adjacent to D.

Given a transport network G, a flow can be indicated in G by labeling each directed edge (i, j) with the ordered pair of numbers $(c(i, j), f(i, j))$. Verify that the flow indicated in Figure 5.4-3 is feasible.

Given a feasible flow f and an edge (i, j), we interpret $f(i, j)$ as signifying that the flow sends $|f(i, j)|$ units of material per unit of time along edge (i, j). If $f(i, j)$ is positive, material flows in the direction assigned to (i, j), but it flows in the opposite direction if $f(i, j) < 0$. Observe that the flow that assigns zero units of material per unit of time to each edge of G is feasible, although otherwise uninteresting.

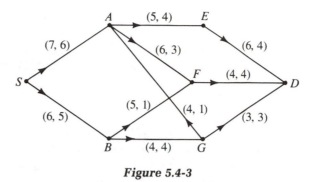

Figure 5.4-3

One method of constructing a maximal feasible flow from S to D starts with an arbitrary feasible flow. (The zero flow can be used, for example.) Then by determining those edges in which the flow is less than the capacity of that edge, we try to augment the total flow from S to D. Observe in Figure 5.4-3 that the flow shown is not maximal, because the route $SAED$ can accommodate an additional flow of one unit. The details of how this general plan can be carried out to produce a maximal feasible flow are somewhat intricate and will be deferred to Section 5.5. The algorithm described there actually constructs a maximal flow in complete detail.

The rest of this section establishes the basic existence theorem for flows in networks, proved by L. R. Ford, Jr., and D. R. Fulkerson in 1956. That theorem provides an interesting way of thinking about the problem, which sometimes leads quickly to a determination of the size of a maximal feasible flow without having to construct such a flow.

Using the network G_1 of Figure 5.4-2 as an example, we imagine that we are interested in stopping completely the flow of material through the network into D. (We can write our own scenario about what material is flowing through G_1 and why we might be interested in stopping that flow.) There are two obvious tactical methods for stopping the flow.

1. Stop the flow at the source by blocking or severing the edges from S to P_i for $i = 1, 2, 3$.

2. Stop the flow at the destination by blocking or severing the edges $D_j D$ for $j = 1, 2, 3, 4, 5$.

Clearly, either of these actions would alter the network, so no material can reach D from S.

Because the sum of the capacities of all edges leaving S is $6 + 17 + 13 = 36$, any feasible flow can deliver at most 36 units of material per unit of time to D. Similarly, the sum of the capacities of edges $D_j D$ for $j = 1$, . . . , 5 turns out to be $7 + 7 + 9 + 3 + 7 = 33$, so any feasible flow can deliver at most 33 units of material per unit of time from S to D. If we could find a particular flow that delivers 33 units per unit of time to D from S, we would know that it is a maximal flow. It is conceivable, however, that although 33 is an upper bound for the value of a feasible flow, no flow with that value is possible. Indeed, consider the set of edges

$$P_3 D_5, \ M_3 D_5, \ M_3 D_4, \ M_3 D_3, \ M_2 D_3, \ M_2 D_2, \ M_2 D_1, \ M_1 D_2, \ M_1 D_1.$$

The sum of the capacities of these nine edges is only 32, and if each of those edges were severed no material could flow from S to D. Hence, the value of a feasible flow is at most 32. As an exercise, we may show that a feasible flow with value 32 does exist in G_1.

To state clearly the central theorem concerning optimal flows in a transport network, we need to define some key terms.

Definition 5.4-3

(a) Let G be a transport network with source S and destination D. A *cut* C of G is a partition of the vertices of G into two subsets X and Y such that X has S as a member and Y (the complement of X) has D as a member.

(b) The *capacity* $c(C)$ of a cut C is the sum of the capacities of all directed edges of the form (x, y), where x is in the partition subset X that contains S and y is in the partition subset Y that contains D.

To justify the word "cut" in this context, suppose that every edge between X and Y were physically severed. Then no material can pass from any vertex in X to any vertex in Y, and, in particular, none can pass from source S to destination D.

Let f denote any feasible flow on G, and let C be a cut of G with capacity $c(C)$. Because any material that flows from S to D must flow along the edges of G that span the cut C, the value of f cannot exceed the capacity of the cut; hence $v(f) \le c(C)$ for every flow f and every cut C. If we now fix f and let C vary over all cuts, we obtain

$$v(f) \le \min_{\text{all } C} c(C).$$

Next let f vary over all feasible flows f on G; we obtain

$$\max_{\text{all } f} v(f) \le \min_{\text{all } C} c(C).$$

In words, *the maximum value of all feasible flows in a transport network can never exceed the minimal value of all possible cuts of the network.*

If this type of argument sounds familiar, it is because we used a similar argument in relation to partially ordered sets in Section 2.5. There we considered two different sets of numbers that are associated with each partially ordered set S:

1. The set of all numbers of chains in the various chain decompositions of S
2. The set of all numbers of elements in the various antichains of S

For each antichain A, each chain decomposition C must have at least as many chains as there are elements in A. Hence, the number c of chains in any chain decomposition C of S and the number a of elements in any antichain A of S must satisfy the inequality $a \le c$. Section 2.5 contained the additional result, known as Dilworth's theorem, that guarantees that $a = c$ for a suitably chosen chain decomposition C and a suitably chosen antichain A.

In any finite partially ordered set S the minimum number of chains in any chain decomposition *equals* the maximum number of elements in any antichain of S.

In the present context we are studying transport networks, and with each network G we can associate two sets of numbers:

1. The set of values of all feasible flows in G
2. The set of capacities of all cuts of G

Each cut C has a capacity $c(C)$ that is at least as large as the value $v(f)$ of any feasible flow f, so the smallest capacity of any cut is at least as large as the largest value of any flow in G.

This means in particular that if we find some flow f' and some cut C' such that $v(f') = c(C')$, then we know that C' is a cut of smallest capacity and f' is a feasible flow of largest value. The fact that equality of these two functions occurs in every transport network for some flow f' and some cut C' is known as the max-flow min-cut theorem. We now state this theorem formally.

Ford – Fulkerson Theorem (Max-Flow Min-Cut Theorem)   In any transport network G let f denote any feasible flow and let C denote any cut. Then the value $v(f)$ of the flow f does not exceed the capacity of the cut C. However, there exists a flow f' and a cut C' such that

$$v(f') = \max_{\text{all } f} v(f) = \min_{\text{all } C} c(C) = c(C').$$

This theorem is the network theory analog of Dilworth's theorem for partially ordered sets; both are members of a growing family of theorems of this type that have arisen in diverse branches of mathematics since the middle of this century.

Exercises 5.4

1. For each of the following five networks
   (a) specify a flow f and record the value $v(f)$ of that flow;
   (b) specify a cut C and record the capacity $c(C)$ of that cut;
   (c) if $v(f) < c(C)$, specify a larger flow or a smaller cut or both, and continue until the value of the flow and the capacity of the cut are equal.

(i)

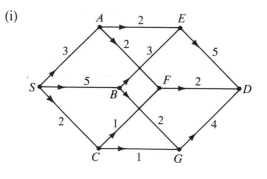

(ii)

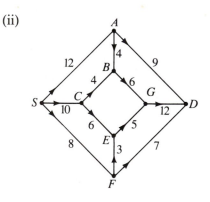

(iii)

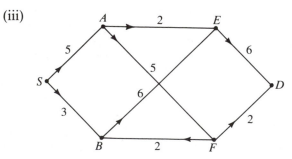

(iv) The network of Figure 5.4-2.

(v)

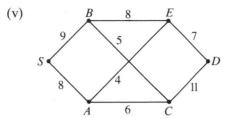

2. Determine a maximum flow from S to D in the following network. Assume that material can flow only in the direction indicated along each edge. State clearly how you are sure that your flow is maximal.

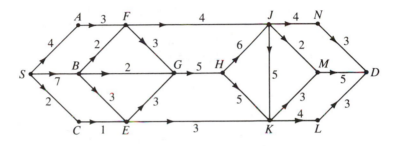

3. Make a list of each cut and its capacity for the graph of
   (a) Exercise 1(iii);
   (b) Exercise 1(v).

4. Describe a maximal feasible flow for the network of Figure 5.4-3. Show that it is indeed maximal.

5.5 Constructing a Maximal Flow

In 1957, within a year after their proof of the max-flow min-cut theorem appeared, Ford and Fulkerson published a second paper that provided an algorithm for constructing a maximal flow in any finite transport network. Many similar algorithms have been published since then, most of which have used the essential ideas and strategies of the original algorithm but have modified the details to improve its computational efficiency.

The basic approach is to start with any flow from source to destination in a given network G and then to search for a way to increase that flow, repeating the augmenting step until no further increase in the flow is possible. By the max-flow min-cut theorem the value $v(f)$ of a flow will be maximal if and only if there is a cut C of G with capacity $c(C)$ such that $v(f) = c(C)$. Thus an effective algorithm should provide a systematic way to increase any flow that is not maximal, together with an automatic signal that no further increase in $v(f)$ is possible.

We first describe such an algorithm informally to explain the overall plan and illustrate it with a simple example. Then we state the algorithm formally and concisely and work some further examples in detail.

Informal Description of a Maximum Flow Algorithm

Consider a transport network G with source S and destination D together with a feasible initial flow f_0 in G with value $v(f_0)$. If f_0 is not a maximal flow, we seek a way to augment f_0 by finding a path from S to D such that each directed edge along that path has a capacity in excess of the present flow through that edge:

$$f_0(x, y) < c(x, y).$$

Having found such a path, we determine the smallest positive excess capacity Δf in all of the edges of that path, and we define a new flow f_1 in G by superimposing on f_0 a flow of Δf along each edge of the path. It follows that

$$v(f_1) = v(f_0) + \Delta f > v(f_0).$$

At this stage we start over, with f_1 as initial flow, and repeat the process of flow augmentation as described in the previous paragraph. We continue repeating until no further flow increment can be found. We need to be certain that the only reason that we can't find a new path to augment the current flow is that no such path exists. To provide such assurance, the algorithm conducts a complete search of all paths, starting from S. It also provides a second confirmation that a maximal flow has been achieved by specifying a cut whose capacity equals the value of the final flow.

EXAMPLE 1   Figure 5.5-1(a) depicts a transport network G that will serve as a simple illustration. Figure 5.5-1(b) shows G with each edge labeled by a pair of integers; the first component is the capacity of that edge, and the second is the value of the present flow along that edge. (The zero flow assigns 0 as the second component for each edge of G, and it is always feasible. But the time needed to determine a maximal flow can be reduced greatly by taking advantage of obvious paths from S to D to define an initial feasible flow f_0 with positive value.) In Figure 5.5-1(b) we have $v(f_0) = 5$.

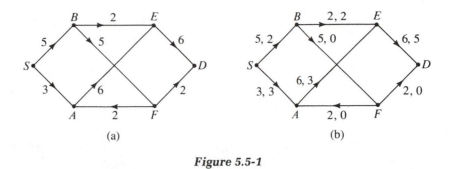

Figure 5.5-1

Referring to Figure 5.5-1(b), we search for a path from S to D, each edge of which has unused capacity. Since (S, A) is already used to capacity, we examine (S, B) and write (S, B) to record that it has unused capacity. We have checked all edges from S, so now we move to the next vertex that has already been examined and found to be at the end of an edge of unused capacity, namely B. Edge (B, E) is fully used, but (B, F) is not, so we write (B, F). Since all edges from B have been looked at, we go on to the next recorded vertex, F, and observe with glee that (F, D) has unused capacity, so we record (F, D). Having reached D from S along edges of unused capacity, we look at our written record $(S, B), (B, F), (F, D)$. Thus a flow-augmenting path turns out to be

$$S, B, F, D$$
$$3, 5, 2$$

where the numbers indicate the unused capacity in each edge of the path. Clearly, a flow increment $\Delta f = 2$ units can be added to each edge of this path to obtain a flow f_1 for which

$$v(f_1) = v(f_0) + 2 = 5 + 2 = 7,$$

as illustrated in Figure 5.5-2(a).

Using f_1 as a new "initial flow," we repeat the search procedure for an augmenting path from S to D. As before, (S, A) has no unused capacity, but (S, B) has. We record (S, B) and move to B. (B, E) has no unused capacity, but (B, F) has. We record (B, F) and move to F. (F, A) has unused capacity, but (F, D) has none, so we record (F, A) and move to A. Now E is the only vertex that is adjacent to A but has not been recorded in this search, and (A, E) has unused capacity. We record (A, E) and move to E, from which we record (E, D) and use the recorded pairs to write a path, working backwards from D to S:

$$D, E, A, F, B, S$$
$$1, 3, 2, 3, 1.$$

Then a flow increment $\Delta f = 1$ can be added along the path $S, B, F, A, E,$ D, as in Figure 5.5-2(b), yielding a flow f_2 for which

$$v(f_2) = v(f_1) + \Delta f = 7 + 1 = 8.$$

From the figure this flow is clearly maximal because all edges from S are used to capacity. Also all edges into D are used to capacity. Either of these conditions is a sufficient (but not necessary) condition for a maximal flow.

But the algorithm asks us to continue to search for an augmenting path. In this case we start at S but there is no adjacent vertex X such that (S, X) has unused capacity, so the search ends with S as the only vertex that has been scanned. Observe that the set $\{S\}$ and its complement $\{A, B, E, F, D\}$ in the set of vertices of G forms a cut C of G, and the capacity of C is the sum of the capacities of all edges directed from a vertex in $\{S\}$ to a vertex in its complement. Thus

$$c(C) = c(S, A) + c(S, B) = 3 + 5 = 8 = v(f_2),$$

which confirms that f_2 is a maximal flow.

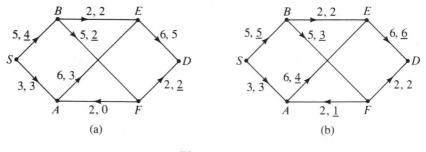

Figure 5.5-2 ∎

Example 1 was a bit too special, because for each vertex V examined in the searches for augmenting flows, each adjacent vertex U either had been examined previously or was the end of a directed edge of the form (V, U) —outgoing from V but never incoming to V. You should notice carefully that both possibilities are accounted for in the formal statement of the maximal flow algorithm. But Example 1 serves well to outline generally the major steps of the algorithm and to give a flow diagram of the algorithm in Figure 5.5-3.

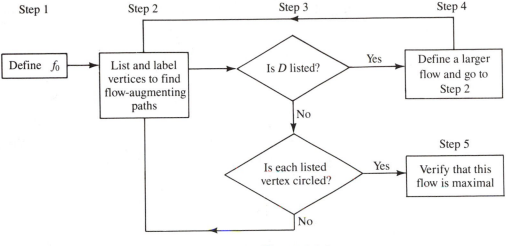

Figure 5.5-3

1. Define an initial flow f_0 on the given network. Choose S as the current vertex and go to **2**.

2, 3. Search systematically for a flow-augmenting path from S to D, keeping careful record of the search, which ends when and only when either D is reached or further progress toward D is impossible. If no further progress is possible, go to **5**. Otherwise go to **4**.

4. Use the record of **2** and **3** to define a new flow, the value of which is larger than the value of the flow at the start of **2**. Then repeat **2** and **3** with the new flow as an initial flow.

5. Having reached **5**, we are assured that the previous applications of **2** and **3** failed to reach D from S. The record of that application defines a cut with capacity equal to the value of the present flow, so the flow must be maximal and the cut must be minimal.

A Maximal Flow Algorithm

Step 1. Given a transport network G with source S and destination D such that each edge of G has an assigned direction, choose any feasible flow f_0 in G as an initial flow and record the value $v(f_0)$. Form a table of two columns with headings Vertex and Label. In the first line of the table enter "S" and "Source" in the first and second columns, respectively.

Step 2. Let x denote the topmost symbol in the vertex column that has not yet been circled. (Originally x will be S, but as the algorithm progresses, x will change.)

(a) For every vertex y that is adjacent to x but not yet listed in the vertex column, if (x, y) is a directed edge having unused flow capacity, write y on the next line of the vertex column and write (x, y) as its label in that same line in the label column.

(b) For every vertex w that is adjacent to x but not yet listed in the vertex column, if (w, x) is a directed edge for which there is a positive flow from w to x, write w on the next line of the vertex column and write $(w, x)*$ as its label. (The asterisk denotes an edge that is directed into x for which there presently is a positive flow into x. Such an edge is called a *reverse edge,* whereas any edge that is directed away from x is called a *forward edge.*)

(c) Never list a vertex that is already listed.

When all edges incident to x have been considered, draw a circle around x on the vertex list.

Step 3. If D does not appear on the vertex list, but some listed vertex remains uncircled, return to step 2; but if D is not on the list, and all listed vertices are circled, go to step 5. If D does appear on the vertex list, go to step 4.

Step 4. The algorithm reaches this step only after D appears on the vertex list. Reading in the label column upward from D, write a succession of adjacent vertices from D to S, and then reverse that succession to obtain a sequence of vertices that defines a path P,

$$P: \quad \{V_0, V_1, V_2, \ldots, V_m\},$$

from $S = V_0$ to $D = V_m$ with unused capacity in each forward edge and positive flow along each reverse edge. For an augmenting flow along the path P to be feasible, its value cannot exceed the unused capacity of any forward edge, nor can it exceed the value of the present flow along any reversed edge.

Denote the present flow in G by f, and let Δf denote the smallest integer in the following two sets of positive integers:

(a) The *unused flow capacities* of all forward-directed edges along P from S to D,

$$c(V_i, V_{i+1}) - f(V_i, V_{i+1})$$

(b) The *present flow* in all reverse-directed edges along P,

$$f(V_{k+1}, V_k)$$

Define a new flow f' on G by this rule; for each edge e of G

$$f'(e) = \begin{cases} f(e) & \text{if } e \text{ is not an edge of } P, \\ f(e) + \Delta f & \text{if } e \text{ is a forward edge of } P, \\ f(e) - \Delta f & \text{if } e \text{ is a reverse edge of } P. \end{cases}$$

Record the value of the new flow. Go to step 2.

Step 5. The algorithm reaches this step if and only if step 3 has been completed, with all vertex symbols in the list being circled and with vertex D *not* on the list. Then the present flow is maximal, and the set of vertices on the list defines a cut C whose capacity is the sum of the capacities of all edges of G that are directed *from* a listed vertex *to* an unlisted vertex. The capacity of C equals the value of the present flow, so that flow is maximal and that cut is minimal.

Before demonstrating the detailed operation of the algorithm in Example 2, we point out some special conditions that must hold in the network G when the algorithm reaches step 5.

1. The search for a flow-augmenting path has failed because no such path exists in G. Some (but not all) vertices have been listed, and each listed vertex has been circled because the list cannot be extended.

2. Let L denote the list of circled vertices, and let L' denote the complement of L in V. The ordered pair $C = \{L, L'\}$ is a partition of V with $S \in L$ and $D \in L'$. Therefore, C is a cut of G.

3. From Definition 5.4-3 the capacity of each cut $C = \{X, Y\}$ is the sum of the capacities of all forward edges across C, a number that varies from cut to cut. However, when the algorithm has reached step 5, the flow across each cut is constant and equal to the maximum flow from S to D. That flow can also be expressed as the sum of the flows along all the forward edges of C, because the flow across each reverse edge of any cut must be 0 when the algorithm reaches step 5.

EXAMPLE 2   Figure 5.5-4(a) shows a variation of the transport network of Example 1, intended to demonstrate how the maximal flow algorithm deals with a reverse edge. Figure 5.5-4(b) shows an initial feasible flow that could be chosen in step 1 of the algorithm. Then the initial operation of the algorithm through steps 2, 3, and 4 is listed. An augmented flow f_1 is shown in Figure 5.5-5. A second application of the algorithm then fails to increase the flow but specifies a cut of minimal capacity, thereby confirming that f_1 is a maximal flow.

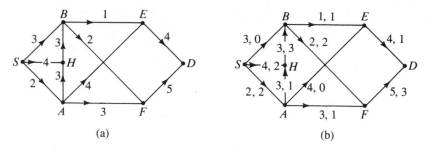

Figure 5.5-4 (a) Network capacities; (b) after step 1: f_0; $v(f_0) = 4$

	Vertex	Label	
1.	Ⓢ	source	Step 2.
2.	Ⓗ	(S, H)	Step 2**a**.
3.	Ⓑ	(S, B)	Step 2**a**. Step 2 is complete.
4.			Step 3. Circle S and apply step 2 to H.
5.	Ⓐ	$(A, H)*$	Step 2**b**; $*$ denotes reverse edge. Step 2 is complete.
6.			Step 3. Circle H and apply step 2 to B.
7.			Step 3. Circle B and apply step 2 to A.
8.	E	(A, E)	Step 2**a**.
9.	F	(A, F)	Step 2**a**. Step 2 is complete.
10.			Step 3. Circle A and apply step 2 to E.
11.	D	(E, D)	Step 2**a**. Step 2 is complete.
12.			Step 3. All vertices are listed, so all vertices will be circled successively. Go to step 4.

By reading the label column from the bottom up, we obtain a backward sequence of edges with unused capacity:

$$(E, D), (A, E), (A, H)*, (S, H),$$

where an asterisk is used to mark each reverse edge. By reversing the order of these edges, we can write a sequence of adjacent vertices from S to D

that describes an augmenting path P. The unused capacity of each edge in P is written just below each vertex pair.

$$S, \quad H, \quad A, \quad E, \quad D$$
$$2 \quad 1 * \quad 4 \quad 3$$

The two sets of numbers cited in step **4a** and **4b** are $\{2, 4, 3\}$ and $\{1\}$, so we let $\Delta f = 1$ and f_1 is obtained by adding 1 to the flow f_0 along edges (S, H), (A, E), and (E, D) and by adding -1 to f_0 along edge (A, H), obtaining the new flow shown in Figure 5.5-5.

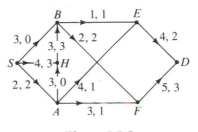

Figure 5.5-5

As specified at the end of step 4, we return to step 2 with a new flow f_1 that we will try to augment.

	Vertex	Label	
1.	Ⓢ	source	Step 2.
2.	Ⓑ	(S, B)	Step **2a**. B and H can be listed in either order.
3.	Ⓗ	(S, H)	Step **2a**. Step 2 is complete.
4.			Step 3. Circle S. Apply step 2 to B.
5.			Step 3. Circle B. Apply step 2 to H.
6.			Step 3. Circle H.

Because no uncircled vertex remains on the list, and D is not on the list, step 3 sends us to step 5. No augmenting path remains, so f_1 is a maximal flow. The set $\{S, B, H\}$ of listed vertices and its complement $\{A, E, F, D\}$ form a cut C of G, and the capacity of C is the sum of the flows along the forward edges (SA, BE, and BF) of the minimal cut C:

$$c(C) = 2 + 1 + 2 = 5 = v(f_1). \qquad ■$$

Exercises 5.5

1. Following the method of Example 2, apply the maximal flow algorithm to each of the following networks, starting with an initial flow as specified. In each case specify a minimal cut.

    (a) The network in Figure 5.5-1(a) with the initial flow given by Figure 5.5-1(b).

    (b) The network in Exercise 5.4-1(ii) with this initial flow:

> 9 from S to A to D
>
> 9 from S to C
>
> 4 from C to B to G
>
> 5 from C to E to G
>
> 9 from G to D
>
> 7 from S to F to D

    (c) The network in Exercise 5.4-1(i) with zero initial flow.

    (d)

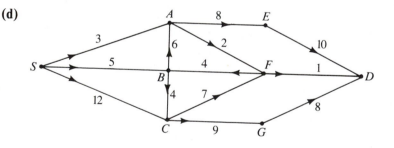

> Initial flow: 8 from S to C to G to D
>
> augmented by 5 from S to B to A to E to D
>
> and also by   3 from S to A to E to D

2. Consider any network G in which the capacity of every edge is the same. Explain how you would determine the value of a maximal flow from source to destination.

3. Determine a maximal flow for the following network and prove that your answer is correct.

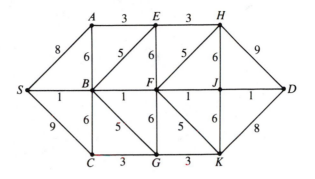

References

Books marked with an asterisk are especially suitable for beginning students; other books are more advanced or specialized.

Aho, A. V., Hopcroft, J. E., and Ullman, J. D. *The Design and Analysis of Computer Algorithms.* Reading, MA: Addison-Wesley, 1974.

* Anderson, Ian. *A First Course in Combinatorial Mathematics.* New York: Oxford University Press, 1979.

Beckenbach, Edwin F. (ed.). *Applied Combinatorial Mathematics.* New York: John Wiley, 1964.

Behzad, M., Chartrand, G., and Lesniak-Foster, L. *Graphs and Digraphs.* Belmont, CA: Wadsworth, 1979.

Berge, Claude. *The Theory of Graphs and its Applications.* New York: John Wiley, 1958.

Berman, Gerald and Fryer, K. D. *Introduction to Combinatorics.* New York: Academic Press, 1972.

Biggs, Norman L., Lloyd, E. Keith, and Wilson, Robin J. *Graph Theory, 1736–1936.* London: Oxford University Press, 1976.

* Bogart, Kenneth P. *Introductory Combinatorics.* Boston: Pitman, 1983.

Bondy, J. A. and Murty, U. S. R. *Graph Theory with Applications.* New York: North-Holland, 1976.

Brualdi, Richard A. *Introductory Combinatorics.* New York: North-Holland, 1977.

Busacker, Robert G. and Saaty, Thomas L. *Finite Graphs and Networks.* New York: McGraw-Hill, 1965.

∗ Chartrand, Gary. *Graphs as Mathematical Models.* Belmont, CA: Wadsworth, 1977.

Christofides, Nicos. *Graph Theory, An Algorithmic Approach.* New York: Academic Press, 1975.

Deo, Narsingh. *Graph Theory with Applications to Engineering and Computer Science.* Englewood Cliffs, NJ: Prentice-Hall, 1974.

∗ Dinkines, Flora. *Introduction to Mathematical Logic.* New York: Meredith, 1964.

Durst, Lincoln K. *The Grammar of Mathematics.* Reading, MA: Addison-Wesley, 1969.

Even, Shimon. *Graph Algorithms.* Potomac, MD: Computer Science Press, 1979.

Goldberg, Samuel. *Introduction to Difference Equations.* New York: John Wiley, 1958.

Hall, Marshall, Jr. *Combinatorial Theory.* Waltham, MA: Blaisdell, 1967.

Harary, Frank. *Graph Theory.* Reading, MA: Addison-Wesley, 1969.

König, Dènes. *Theorie der Endlichen und Unendlichen Graphen.* New York: Chelsea, 1950.

Levy, Leon S. *Discrete Structures of Computer Science.* New York: John Wiley, 1980.

∗ Liu, C. L. *Elements of Discrete Mathematics.* New York: McGraw-Hill, 1977.

Liu, C. L. *Introduction to Combinatorial Mathematics.* New York: McGraw-Hill, 1968.

∗ Niven, Ivan. *Mathematics of Choice.* Washington, DC: The Mathematical Association of America, 1965.

∗ Ore, Oystein. *Graphs and Their Uses.* New York: Random House, 1963.

Pólya, George, Tarjan, Robert E., and Woods, Donald R. *Notes on Introductory Combinatorics.* Boston: Birkhaüser, 1983.

Prather, Ronald E. *Discrete Mathematical Structures for Computer Science.* Boston: Houghton Mifflin, 1976.

Riordan, John. *An Introduction to Combinatorial Analysis.* New York: John Wiley, 1958.

∗ Roberts, Fred S. *Applied Combinatorics.* Englewood Cliffs, NJ: Prentice-Hall, 1984.

Roberts, Fred S. *Graph Theory and Its Applications to Problems of Society.* Philadelphia: Society for Industrial and Applied Mathematics, 1978.

* Ross, Kenneth A. and Wright, Charles R. B. *Discrete Mathematics.* Englewood Cliffs, NJ: Prentice-Hall, 1985.

Stanat, Donald and McAllister, David F. *Discrete Mathematics in Computer Science.* Englewood Cliffs, NJ: Prentice-Hall, 1977.

* Stoll, Robert R. *Sets, Logic, and Axiomatic Theories.* New York: W. H. Freeman, 1961.

Tucker, Alan. *Applied Combinatorics.* New York: John Wiley, 1980.

Vilenkin, N. Ya. *Combinatorics.* New York: Academic Press, 1971.

Whitworth, William A. *Choice and Chance.* London: George Bell and Sons, 1878.

Answers to
Selected Exercises

Section 1.1

1. There are only two Graeco or Latin squares of order two:

$$\begin{matrix} A & B \\ B & A \end{matrix} \quad \text{and} \quad \begin{matrix} B & A \\ A & B \end{matrix} \ .$$

Their superimposition contains the ordered pair AB twice, so they are not orthogonal.

3. $\begin{matrix} Ad & Ba & Cb & Dc \\ Bc & Ab & Da & Cd \\ Ca & Dd & Ac & Bb \\ Db & Cc & Bd & Aa \end{matrix}$

Section 1.2

1. (a) The set of all natural numbers larger than 15 that are evenly divisible by 3.

  (b) The set of integers 3 and -2.

  (c) The set whose members are the natural numbers 3 and 4 and the 2-element set whose members are the natural numbers 1 and 2.

3. (a) $\{x \in N \mid x \neq 0 \text{ and } x < 13\}$.

(b) $\{x \in N \mid x = 2^n \text{ for some } n \in N\}$.

5. $A = \{-3, -2, -1, 0, 1, 2, 3\}$, so $A = E$. Thus $A \subseteq E$ and $E \subseteq A$. Also $X \subseteq X$ for all sets $X = A, B, C, D, E$. $B \subseteq A$, $B \subseteq E$, and $B \subseteq C$. D is the empty set, so $D \subseteq A$, $D \subseteq B$, $D \subseteq C$, and $D \subseteq E$.

7. Each subset of $\{1, 5, 10, 25\}$ has a different sum; there are $16 = 2^4$ subsets of a 4-element set, so there are 16 different sums of money.

9. (a) Suppose $S \in U$ and S is weird—that is, $S \in S$. By definition S is the set of all nonweird sets, so S is not weird.

(b) Conversely if S is not weird, by definition $S \in S$, so S is weird.

Section 1.3

1. (a) $B \cap C = \{1\}$.
$A \cup (B \cap C) = \{1, 2, 3, 4, 5, 6\}$.
$A \cup B = \{1, 2, 3, 4, 5, 6, 7\}$.
$A \cup C = \{0, 1, 2, 3, 4, 5, 6\}$.
$(A \cup B) \cap (A \cup C) = \{1, 2, 3, 4, 5, 6\}$.

(b) $B \cup C = \{0, 1, 2, 6, 7\}$.
$A \cap (B \cup C) = \{2, 6\}$.
$A \cap B = \{6\}$.
$A \cap C = \{2\}$.
$(A \cap B) \cup (A \cap C) = \{2, 6\}$.

5.

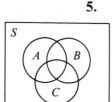 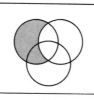 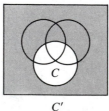 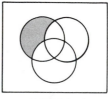

$A \cap B'$ C' $(A \cap B') \cap C'$

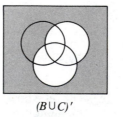 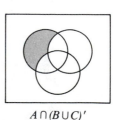

$(B \cup C)'$ $A \cap (B \cup C)'$

The two final drawings (one on each line) are identical.

7. (a) $A \cap B = \varnothing$.   0 elements
$A \cap C = \{i, t\}$.   2 elements
$A \cup B = \{a, d, e, g, i, l, m, n, r, t, u\}$.   11 elements
$A \cup C = \{a, e, i, l, m, n, o, r, t, u\}$.   10 elements

(b) No, not if the same letter appears in the two sets.

Section 1.4

1. (a) $n(D \cup C) = 50$.

   (b) $n(D \times C) = 600$.

   (c) $n(D \times (C \cup T)) = 780$.

   (d) $n(D \times C \times T) = 5400$.

3. (a) $n((A \cup B) \times C) = (k + m)(p)$.

   (b) If $n[A \cap (B \times C)] = q$, then
   $$n[A \cup (B \times C)] = n(A) + n(B \times C) - n[A \cap (B \times C)]$$
   $$= k + mp - q.$$
   The value of q must be known.

   (c) $n(A \cup A) = n(A) = k$.
   $n(B \times B) = m^2$.
   $n[(A \cup C) \times (B \cup C)] = (k + p)(m + p)$.
   $n(A \times B \times C) = n(A)n(B)n(C) = kmp$.

5. 60,000.

7. $1 + 7 + 7^2 + 7^3 + 7^4$ (man, wives, sacks, cats, kits) $= 2801$. At least one person was going *to* St. Ives.

9. (a) Let $S = \{1, 2, 3\}$. The subsets of S are $\varnothing$; $\{1\}$, $\{2\}$, $\{3\}$; $\{1, 2\}$, $\{1, 3\}$, $\{2, 3\}$; $\{1, 2, 3\}$.

   (b) Let $T = S \cup \{4\}$. The subsets of T are the subsets of S and the sets obtained by adjoining 4 as an element to each subset of S. Hence T has twice as many subsets as S.

   (c) The number of subsets of an n-element set is 2^n.

11. $9 \cdot 10 \cdot 10 \cdot 9 = 8100$.

Section 1.5

1. (a) 8.

   (b) 32.

3. Five respondents checked Bach and Debussy but not Copland.

5. (b) 171 patients.

7.

Sex Age Employment	Female *F*			Male *F'*			Combined *F ∪ F'*		
	A	*A'*	**Sum**	*A*	*A'*	**Sum**	*A*	*A'*	**Sum**
E	197	0	197	301	0	301	498	0	498
E'	62	172	234	70	170	240	132	342	474
Sum	259	172	431	371	170	541	630	342	972

11. $1050 - [350 + 210 - 70] = 1050 - 490 = 560.$

Section 1.6

1. (a) $p \wedge \sim q.$

   (c) $(\sim p) \wedge (\sim q).$

   (g) $\sim(p \wedge q).$

3. a, b, d, f, and **h** are statements. **c, e,** and **g** are not statements.

5. (b)

p	*q*	*p ∧ q*	*~(p ∧ q)*	*~(p ∧ q) ∧ (p ∨ q)*
1	1	1	0	0
1	0	0	1	1
0	1	0	1	1
0	0	0	1	0

   (c)

p	*q*	*p ∨ q*	*~(p ∨ q)*	*[~(p ∨ q)] ∧ p*
1	1	1	0	0
1	0	1	0	0
0	1	1	0	0
0	0	0	1	0

7. (a) I read a book, and the sun is not shining.

   (c) I either read a book or play tennis, but I do not both read a book and play tennis.

   (e) Either the sun is shining or I read a book, or the sun is not shining and I do not play tennis.

Section 1.7

1.

p	q	$q \vee \sim p$	$(\sim q) \vee p$	(a) $(q \vee \sim p) \wedge [(\sim q) \vee p]$	(b) $p \Leftrightarrow q$	(c) $[p \vee (\sim q)] \Rightarrow \sim p$
1	1	1	1	1	1	0
1	0	0	1	0	0	0
0	1	1	0	0	0	1
0	0	1	1	1	1	1

p	q	$\sim p$	$(p \Rightarrow \sim q)$	$(\sim p) \Rightarrow (p \Rightarrow \sim q)$	$(p \Rightarrow \sim q) \Rightarrow \sim p$	(d) $(\sim p) \Leftrightarrow (p \Rightarrow \sim q)$
1	1	0	0	1	1	1
1	0	0	1	1	0	0
0	1	1	1	1	1	1
0	0	1	1	1	1	1

2.

Statement	*Converse*	*Contrapositive*
(b) $p \Rightarrow (q \Rightarrow r)$	$(q \Rightarrow r) \Rightarrow p$	$\sim (q \Rightarrow r) \Rightarrow \sim p$

3. (a) If p is valid, then so is "p or q."

   (d) "p implies q" is equivalent to "not q implies not p."

5. (a) $[(p \wedge q) \Rightarrow p]$
$$\equiv p \vee [\sim (p \wedge q)] \equiv p \vee (\sim p \vee \sim q) \equiv (p \vee \sim p) \vee \sim q$$
$$\equiv u \vee \sim q \equiv u. \text{ Hence, it is a tautology.}$$

7. The first statement is the converse of the given statement, and the second is the contrapositive.

   (a) If an integer is the fourth power of some integer, then it is the square of some integer. True.
   If an integer is not the fourth power of some integer, then it is not the square of some integer. False.

   (b) If at least one of two positive integers is even, then their product is even. True.
   If two positive integers are both odd, then their product is odd. True.

   (d) If at least one of two positive integers is greater than 9, then their sum is 20. False.
   If each of two positive integers is less than or equal to 9, then their sum is not equal to 20. True.

Section 1.8

5. (b)

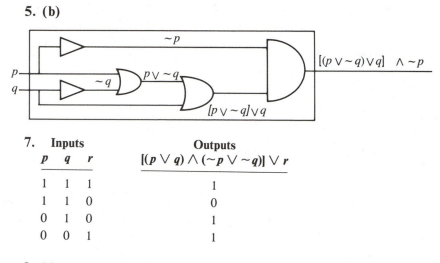

7.

Inputs			Outputs
p	**q**	**r**	$[(p \vee q) \wedge (\sim p \vee \sim q)] \vee r$
1	1	1	1
1	1	0	0
0	1	0	1
0	0	1	1

8. (c)

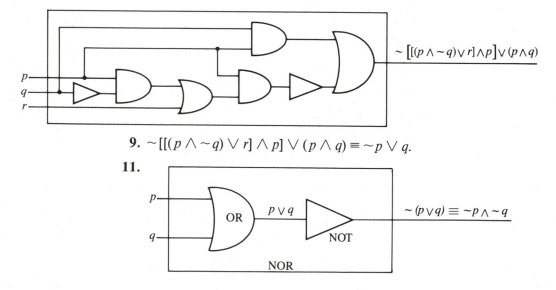

9. $\sim [[(p \wedge \sim q) \vee r] \wedge p] \vee (p \wedge q) \equiv \sim p \vee q.$

11.

Section 1.9

1. (a) 3406.057.

  (b) 0.050007.

4. (b) 11001011.

  (d) 0.101.

5. (b) 21.

    (d) 5.125.

6. (b) 1101011.

    (d) 1010100011.

12. If m denotes a smallest positive rational number, then $m > 0$ and $m > m/2 > 0$. Therefore $m/2$ is a positive rational number smaller than m, which contradicts the definition of m, so a smallest positive rational number does not exist.

15. (a) No.

    (b) Yes.

    (c) The set of irrational numbers is *not* closed under either addition or multiplication.

16. (a) $\lhd$ is irreflexive, asymmetric, and transitive. Also trichotomy holds in "appears to the left of."

17. Base 8

    (b) 1211.

    (d) 0.5.

18. Base 10

    (b) 567.

    (d) 7.3125.

20. (a) 201.

    (d) 6366.

Section 1.10

1. (a) Prove: $0 + 1 + \cdots + k = k(k + 1)/2$ (by induction).
Let $S =$ set of all k for which this assertion is true.

Step 1. If $k = 0$, the statement is $0 = 0(1)/2$, which is true, so $0 \in S$.

Step 2. Suppose $0 + 1 + \cdots + m = m(m + 1)/2$. Then

$$(0 + 1 + \cdots + m) + (m + 1) = \frac{m(m + 1)}{2} + (m + 1)$$

$$= \frac{m(m + 1) + 2(m + 1)}{2} = \frac{(m + 1)(m + 2)}{2}.$$

Hence if $m \in S$, then $m + 1 \in S$.

Step 3. By the principle of induction $S = N$.

5. Prove for $n > 0$: $1^2 + 3^2 + 5^2 + \cdots + (2n - 1)^2 = \dfrac{(4n^3 - n)}{3}$.

   Let $S =$ set of all $n \in N$ such that the statement is true.

   Step 1. $1^2 = (4 \cdot 1^3 - 1)/3$, so $1 \in S$.
   Step 2. Let $m \in S$. Then $1^2 + 3^2 + \cdots + (2m - 1)^2 = (4m^3 - m)/3$.
   Then

   $$[1^2 + 3^2 + \cdots + (2m - 1)^2] + (2m + 1)^2 = \frac{4m^3 - m}{3}$$

   $$+ \frac{12m^2 + 12m + 3}{3} = \frac{4(m^3 + 3m^2 + 3m + 1) - (m + 1)}{3}$$

   $$= \frac{4(m + 1)^3 - (m + 1)}{3}.$$

   So $m + 1 \in S$ when $m \in S$.
   Step 3. By induction, statement is valid for all $n \geq 1$.

10. **Step 1.** The assertion is valid for $k = 5$; $2^5 > 5^2$.
    Step 2. Assume that the assertion is valid for $m > 5$.

    $$2^{m+1} = 2 \cdot 2^m > 2(m^2) = m^2 + m^2 \geq m^2 + (2m + 1) = (m + 1)^2.$$

    The last statement holds for all $m > 2$—that is, $m^2 \geq 2m + 1$ when $m > 2$.
    Step 3. By induction $2^n > n^2$ for all $n > 5$.

Section 2.1

1. (a) The relation is irreflexive. It does not possess any of the remaining five properties.

   (e) The relation is reflexive and symmetric. It does not possess any of the remaining four properties.

5. There are 2^3 subsets of $A \times B$, so there are 8 binary relations from A to B. One is $R = \{(0, 2)\}$.

6. (a) $R(1)$ is reflexive, symmetric, and transitive. It does not possess any of the remaining three properties.

(d) $R(4)$ is reflexive, antisymmetric, and transitive. It does not possess any of the remaining three properties.

9. (b) Must adjoin $(0, 2)$, $(1, 1)$, $(2, 2)$, $(1, 3)$, and $(0, 3)$.

10. R is symmetric: Let $(a, b) R (c, d)$. Then $a + d = b + c$. Hence, $c + b = d + a$, so $(c, d) R (a, b)$.

Section 2.2

1. (a) R consists of the ordered pairs (A, F), (A, B), (B, A), (B, C), (C, E), (D, B), (E, B), (F, C), (F, D).

(b) R is not transitive.

(c)

	A	B	C	D	E	F
A	0	1	0	0	0	1
B	1	0	1	0	0	0
C	0	0	0	0	1	0
D	0	1	0	0	0	0
E	0	1	0	0	0	0
F	0	0	1	1	0	0

4. (a) R consists of the ordered pairs
(A, B), (A, D), (A, F)
(B, D), (B, E), (B, F)
(C, D), (C, F)
(D, A), (D, B)
(E, A), (E, C)
(F, E)

(b)

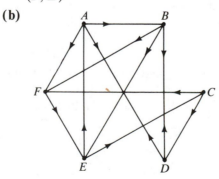

7. (a)

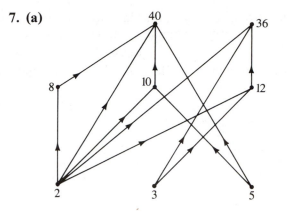

(b)

	2	3	5	8	10	12	36	40
2	0	0	0	1	1	1	1	1
3	0	0	0	0	0	1	1	0
5	0	0	0	0	1	0	0	1
8	0	0	0	0	0	0	0	1
10	0	0	0	0	0	0	0	1
12	0	0	0	0	0	0	1	0
36	0	0	0	0	0	0	0	0
40	0	0	0	0	0	0	0	0

10. (a)

	1	2	3	4	5
1	0	1	1	1	0
2	0	0	1	1	0
3	0	0	0	1	1
4	0	1	0	0	0
5	1	1	0	1	0

(c) Persons 1, 3, and 5 can reach all others in no more than two stages.

1 to 2, 3, 4 directly and to 5 through 3.

3 to 4, 5 directly $\begin{cases} \text{and to 1 through 5,} \\ \text{and to 2 through 4 or 5.} \end{cases}$

5 to 1, 2, 4 directly and to 3 through 1 or 2.

12. (a)

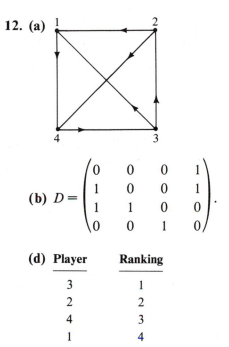

(b) $D = \begin{pmatrix} 0 & 0 & 0 & 1 \\ 1 & 0 & 0 & 1 \\ 1 & 1 & 0 & 0 \\ 0 & 0 & 1 & 0 \end{pmatrix}.$

(d)

Player	Ranking
3	1
2	2
4	3
1	4

Section 2.3

1. (a) $2A = \begin{pmatrix} 4 & 0 & 2 \\ -2 & 6 & 0 \\ 2 & 2 & 4 \end{pmatrix}.$

(d) $BA = \begin{pmatrix} 8 & -8 & 4 \\ 1 & 3 & 1 \\ -3 & 5 & -2 \end{pmatrix}.$

3. (a) $\sum_{i=1}^{n+1} i^3$ or $\sum_{i=0}^{n} (i+1)^3.$ **(d)** $\sum_{k=1}^{5} a(k, i)b(j, k).$

5. (b) $M + M^2 = \begin{pmatrix} 0 & 2 & 2 & 2 & 1 & 1 & 2 \\ 0 & 1 & 1 & 4 & 1 & 3 & 3 \\ 0 & 1 & 1 & 4 & 1 & 2 & 2 \\ 0 & 0 & 0 & 0 & 0 & 0 & 0 \\ 0 & 0 & 0 & 3 & 0 & 2 & 1 \\ 0 & 0 & 0 & 1 & 0 & 0 & 0 \\ 0 & 0 & 0 & 2 & 0 & 1 & 0 \end{pmatrix}.$

7. (a) $M + M^2 = \begin{pmatrix} 0 & 2 & 2 & 3 & 1 \\ 0 & 1 & 1 & 2 & 1 \\ 1 & 2 & 0 & 2 & 1 \\ 0 & 1 & 1 & 1 & 0 \\ 1 & 3 & 2 & 3 & 0 \end{pmatrix}.$

(c) $P(5)$.

11. (a) The (i, j) entry of J^3 gives the number of ways $P(i)$ can influence $P(j)$ through two intermediaries.

Section 2.4

2. (a) An equivalence relation.

(b) Not an equivalence relation.

3. (a) Concentric circles with center at p and radii $r \geq 0$.

5. (a) $a \not{R} a$ for some a.

8. (a) R is an equivalence relation on I.

(b) R is not an equivalence relation on I.

9. (a) $[1]$ is the set of all odd integers.

10. (b) $[\frac{2}{3}]_E = \{x \in Q \mid x = \frac{2}{3} + k \text{ for some } k \in I\}$.
$[401]_E = I$.

11. (c) $[B]_E = \{B, D, G\}$ is one of the 4 E-equivalence classes of S.

Section 2.5

2. R is not a partial ordering of $\mathcal{F}$ if S has more than one element.

3. (b) D is not a total ordering of S.

(c)

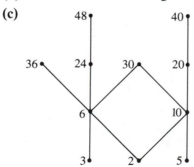

5. **(a)** There are 24 chains of three or more elements.

   (b) There are 21 antichains containing three or more elements.

   (c) There are many examples of five-chain decompositions. For example, {36, 6, 3}, {48, 24}, {30, 10, 5}, {2}, {40, 20}.

6. A maximal antichain in Figure 2.5-3 is {16, 12, 18, 20, 14, 15, 11, 13, 17, 19}, which contains 10 elements.
   A minimal chain decomposition, having 10 chains, is given by {16, 8, 4, 2, 1}, {12, 6, 3}, {18, 9}, {20, 10, 5}, {14, 7}, {15}, {11}, {13}, {17}, {19}.

8. R is a strong total ordering on A.

Section 2.6

1. **(a)** $f(0) = -2$.
   $f(-\frac{1}{2}) = -\frac{7}{2}$.
   $f(\frac{1}{2}) = -\frac{1}{2}$.

   (b) *Hint:* Show that if $f(a) = f(b)$, then $a = b$.

   (c) *Hint:* Begin with an arbitrary real number b and show that there is a real number a such that $f(a) = b$.

3. **(a)** $(g \circ f)(-1) = 26$.

   (b) $(f \circ g)(-1) = 4$.

   (c) $(g \circ f)(x) = 9x^2 - 12x + 5$,
   $(f \circ g)(x) = 3x^2 + 1$.

5. **(d)** Let c be any real number; then $f(d) = c$ for $d = (c - 1)/2$.

   (e) Let $h(x) = (x - 1)/2$. Then

   $$(f \circ h)(x) = 2\left(\frac{x - 1}{2}\right) + 1 = x \quad \text{and}$$

   $$(h \circ f)(x) = \frac{(2x + 1) - 1}{2} = x.$$

   The function h is the inverse of f.

7. **(a)** f and g are one-to-one.
   Neither f nor g is onto N.

   (b) F is not one-to-one, but G is one-to-one.
   F is not onto I, but G is onto I.

9. **(a)** $f(\frac{10}{3}) = 3, f(-\frac{10}{3}) = -4, f(3) = 3, f(-3) = -3$.

   (b) f is onto I.

(d) The function g has the set $\{0, 1\}$ as its range.

11. (a) f, h are one-to-one.

(b)

x	$f^{-1}(x)$
a	1
b	2
c	3
d	4

x	$h^{-1}(x)$
2	a
3	b
4	c

14. (b) M has at most one nonzero entry in each column.

17. $S = \{a, b, c, d, r, s, t\}$,
$A = \{a, b, r, s\}$, $B = \{b, s, t\}$.

x	$I_A(x)$	$I_B(x)$	$I_{A'}(x)$	$I_{A \cap B}(x)$	$I_{A \cup B}(x)$
a	1	0	0	0	1
b	1	1	0	1	1
c	0	0	1	0	0
d	0	0	1	0	0
r	1	0	0	0	1
s	1	1	0	1	1
t	0	1	1	0	1

From this table each of the following formulas can be verified for all x in S:

$$I_{A'}(x) = 1 - I_A(x)$$
$$I_{A \cap B}(x) = I_A(x) I_B(x)$$
$$I_{A \cup B}(x) = I_A(x) + I_B(x) - I_{A \cap B}(x)$$

Also the sum of the second column is 4, the number of elements in A; and the sum of the third column is 3, the number of elements in B.

Section 2.7

1. (a) $\{M; \blacktriangleleft\}$ is a partially ordered set.

(b) $\{M; \blacktriangleleft\}$ is not totally ordered.

3. (b) $*$ is commutative.

(c) $*$ is not associative.

    (d) ▶ is not a partial ordering.

5. (a) $\{N; +\}$ is a monoid but not a group.

    (b) $\{I; +\}$ is a group.

7. (b) $p \oplus q = 2 + 2t + 3t^2 + 4t^3 + 5t^4 + 6t^5 + \cdots$,
        $p \odot q = 1 + 2t + 4t^2 + 7t^3 + 11t^4 + 16t^5 + \cdots$.

Section 3.1

1. $F(13) = 233$.

3. 738.

4. *Hint:* Join the midpoints of the sides of the triangle, thus subdividing the triangle into four congruent small triangles. Use the pigeonhole principle.

7. *Hint:* The sum of the numbers 1 through 12 is $12(13)/2 = 78$. There are 12 sets of 5 adjacent numbers, and in those 12 sets each number appears 5 times; hence the total sum is $78(5) = 390$. Use the pigeonhole principle.

9. $P(2 \text{ or } 3 \text{ at least once in 6 rolls}) = 665/729$.

11. $P(3) = 24/55$, $P(4) = 16/55$.

13. The minimum number of slices is six.

Section 3.2

2. (a) $(14!)(12!)$ ways.

    (b) $(14)(12) = 168$ handshakes.

4. 2520.

5. (a) 720.

    (c) 240.

7. $[(213)(212)(211) \cdots (199)] = P(213, 198)$.

9. 4536.

11. (a) 43,680.

    (b) 5280.

Section 3.3

1. (a) $C(52, 13) = \dfrac{52!}{39! \; 13!}$.

    (b) $\text{Prob}(4, 3, 3, 3) = \dfrac{C(4, 1)C(13, 4)C(13, 3)^3}{C(52, 13)}$.

    (c) $\text{Prob}(2 \text{ aces}) = \dfrac{C(4, 2)C(48, 11)}{C(52, 13)}$.

3. $C(40, 20)C(30, 25)C(30, 5)$.

4. (d) $80 : 9 : 4$.

5. The two products are equal.

8. (a) $C(n + 1, k) = \dfrac{(n + 1)!}{(n + 1 - k)! \; k!} = \dfrac{(n + 1)}{(n + 1 - k)} \dfrac{n!}{(n - k)! \; k!}$

$$= \dfrac{n + 1}{(n + 1 - k)} C(n, k).$$

11. $2^n = (1 + 1)^n = \displaystyle\sum_{k=0}^{n} C(n, k) \cdot 1^{n-k} \cdot 1^k = C(n, 0) + C(n, 1) +$

$$\cdots + C(n, n).$$

The number of subsets of all sizes of an n-element set is 2^n.

13. *Hint:* $\displaystyle\sum_{k=0}^{n} C(n, k)(-1)^k = 0$.

$$[C(n, 0) + C(n, 2) + C(n, 4) + \cdots + C(n, m)]$$
$$- [C(n, 1) + C(n, 3) + \cdots + C(n, k)] = 0,$$

where $m = n, k = n - 1$ if n is even, and $m = n - 1, k = n$ if n is odd.

Section 3.4

1. $C(30; 9, 9, 5, 5, 1, 1) = C(30, 9)C(21, 9)C(12, 5)C(7, 5)C(2, 1)$

$$= \dfrac{30!}{9! \; 9! \; 5! \; 5!}.$$

3. 83,160.

6. (a) $C(3, 1)C(5, 1)C(7, 3) + C(3, 2)C(5, 0)C(7, 3) +$
      $C(3, 1)C(5, 0)C(7, 4) = 735$.

    (b) 2156.

7. 1,961,256.

9. (a) $\dfrac{16!}{5!\,4!\,2!\,2!}$.

(b) $\dfrac{15!}{5!\,3!\,2!\,2!} + \dfrac{15!(2)}{5!\,4!\,2!\,2!}$.

11. 59 divisors greater than 1.
(*Hint:* Any divisor $n > 1$ of 10,800 has the form $n = 2^a 3^b 5^c$, where $0 \le a \le 4$, $0 \le b \le 3$, $0 \le c \le 2$, and $a + b + c > 0$.)

Section 3.5

1. (a) 3^{10}.

(b) $3(2^9)$.

3. (a) 126.

(b) 495.

(c) $9^4 = 6561$.

5. 180,000.

7. $C(7 + 3 - 1, 3 - 1) = 36$.

9. (a) 420.

(b) 420.

11. $(8!) \cdot 7^3$.

Section 3.6

1. $C(10 + 5 - 1, 10) = 1001$.

3. $C(46 + 4 - 1, 46) = C(49, 3)$.
(*Hint:* The number of different solutions in positive integers of the inequality $x_1 + x_2 + x_3 < 50$ is equal to the number of different solutions in *positive integers* of the equation $x_1 + x_2 + x_3 + x_4 = 50$.)

5. $6^4 = 1296$.

7. $C(10 + 5 - 1, 10) = 1001$.

9. $C(5 + 5 - 1, 4) = 126$.

12. (a) $C(31 + 3 - 1, 31) = 528$.

Section 3.7

1. (a) $P(9, 9)C(8, 3) = 9!(56)$.
   (b) $9![C(12, 3) - C(8, 3)] = 9!(164)$.

3. We interpret this problem as an assignment problem and use Theorem 3.6-4. If the order in which the cans of food are handed out is considered important, then it is an arrangement problem and we would use Theorem 3.7-1. With our interpretation there are 5^{32} ways to parcel out the food.

5. (a) $(6!)7 = 7!$.
   (b) $(6!)4 = 2880$.

7. (a) $11 \cdot 7!$.
   (b) $3 \cdot 7!$.

9. (a) $24!C(9 + 2, 2) = 24!(55)$.
   (b) $24!(127)$.

Section 3.8

1. The coefficient of $x^d y^e z^f$ in $(x + y + z)^n$ is the number of ways in which x can be chosen from d of the n factors $(x + y + z)$ and in which y can be chosen from e of the remaining $n - d$ factors, leaving $f = n - e - d$ factors from which z must be chosen.

$$T(n; d, e, f) = C(n, d)C(n - d, e)C(n - d - e, f)$$

$$= \frac{n!}{d!(n-d)!} \cdot \frac{(n-d)!}{e!(n-d-e)!} \cdot \frac{(n-d-e)!}{f!(n-d-e-f)!}$$

$$= \frac{n!}{d! \, e! \, f!}.$$

3. For all k such that $0 < k \le n$,

$$T(n; k, n - k, 0) = T(n; k, 0, n - k) = T(n; 0, k, n - k) = C(n, k).$$

5. $s(n) = s(n - 1) + s(n - 2)$ for $n > 2$; $s(1) = 2$, $s(2) = 3$.
   $s(8) = 55$.

8. (a) $S(n) = \sum_{k=1}^{n} S(k - 1)S(n - k)$.
   (b) $S(5) = 42$.

Section 3.9

1. **(a)** $g(x) = (x^2 + x^3 + x^4 + x^5)(1 + x + x^2 + x^3)^2$
$= x^2(1 + x + x^2 + x^3)^3$.

(b) 10.

3. $g(x) = (1 + x^5 + x^{10} + \cdots + x^{100})$
$\cdot (1 + x^{10} + x^{20} + \cdots + x^{100})(1 + x^{25} + x^{50} + x^{75} + x^{100})$.
The desired number is the coefficient of x^{100}.

5. **(a)** $g(x) = (1 + x^2 + x^4 + x^6 + x^8)(1 + x^3 + x^6 + x^9)(1 + x^5 + x^{10})$.

(c) $\Sigma_{k=0}^{\infty} a_k = 60$. The total number of different ways of making post-age of some amount with four 2-cent stamps, three 3-cent stamps, and two 5-cent stamps is 60, and the amounts are all numbers from 0 through 27 except for 1 and 26.

7. **(a)** $g(x) = (x^2 + x^3 + \cdots)^2(x^2 + x^4 + x^6 + \cdots) = \sum_{k=6}^{\infty} a_k x^k$
$= x^6(1 + x + x^2 + \cdots)^2(1 + x^2 + x^4 + \cdots)$.

(b) $a_{12} = 16$.

9. $g(x) = (1 + x + x^2 + x^3 + x^4)(1 + x + x^2 + x^3 + x^4 + x^5)$
$\cdot (1 + x + x^2 + x^3)$; 120 collections of fruit.

Section 3.10

1. $g(x) = x^2(1 + x + x^2 + x^3 + \cdots)^3 = x^2(1 - x)^{-3}$
$= x^2[C(2, 0) + C(3, 1)x + C(4, 2)x^2 + C(5, 3)x^3 + \cdots]$.

2. **(b)** $C(23, 18) - C(14, 9) = 31{,}647$.

3. 10.

5. $g(x) = (1 + x)^{15}(1 + x^2 + x^5)$; coefficient of $x^{13} = 7905$.

7. $g(x) = (x + x^2 + x^3 + \cdots)^5 = x^5(1 - x)^{-5}$;
coefficient of $x^{15} = 1001$.

Section 4.1

1. (a)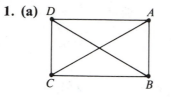

(b) Paths from A to B are AB, ACB, ADB, $ACDB$, $ADCB$.

3. (a)

n	Edges in K_n
2	1
3	3
4	6
5	10
6	15

5. (a) The complement of $K_{3,3}$ is $K_3 \cup K_3$:

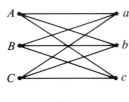 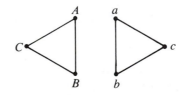

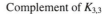

$K_{3,3}$ Complement of $K_{3,3}$

Section 4.2

2. (c)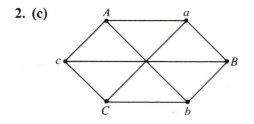

4. (a), **(b)**, and **(c)** are isomorphic; **(d)** is not isomorphic to any of the others.

6. (a) and **(b)** are isomorphic; **(c)** is not isomorphic to **(a)** or **(b)**.

8. (b) No.

10. There are three nonisomorphic graphs.

Section 4.3

6. $e \le 5(v - 2)/3$.

7. If $v < 5$, that component cannot contain K_5 or $K_{3,3}$ within it; hence it is planar.

9. (b) Planar; **(e)** not planar.

10. (c)

m	*n*	*f*	*e*	*v*	Shape of Face	
3	3	4	6	4	Triangle	(Tetrahedron)
3	4	6	12	8	Square	(Cube)
3	5	12	30	20	Pentagon	(Dodecahedron)
3	6	No solution			Triangle	—
4	3	8	12	6	Triangle	(Octahedron)
4	4	No solution			Square	—
5	3	20	30	12	Triangle	(Icosohedron)
6	3	No solution			Triangle	—

Section 4.4

1. (a) An Euler circuit exists but no Euler trail.

(b) An Euler trail exists but no Euler circuit.

(g) An Euler trail exists but no Euler circuit.

3. A separable graph has an Euler circuit if and only if $\deg(v) > 2$ and all vertices have even degree. (The vertex v is a vertex of separability.)

5. Decimal form of a sequence: 0, 1, 2, 4, 9, 3, 7, 15, 14, 13, 10, 5, 11, 6, 12, 8. (There are other correct sequences.)

9. It suffices to close 2 bridges.

Section 4.5

2. (a)

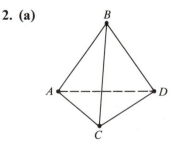

A Hamilton circuit is *ABCDA*.

4. There is a Hamilton circuit.

6. (a) *BCPNMDFKLTSRQZXWVJHGB* is a Hamilton circuit with the given start.

9. (b)

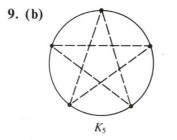

K_5

Let $n = 5$. The $(n - 1)/2 = 2$ Hamilton circuits in the figure are two disjoint sets of edges of K_5.

10. Five meetings are possible.

Section 4.6

1. (a) The chromatic numbers are 3 and 4, respectively.

(c) The chromatic number is 4 for each of the graphs.

5. Two meeting times are needed.

6. (b) The edge-chromatic number is 5.

9. (a) Three days are required.

(b) Five days are required.

10. Twelve gondolas of sand.

Section 4.7

1. (a) 82.

(b) 7.

4. Having been assigned B as a starting vertex, proceed to step 2 of Prim's algorithm. Choose a shortest edge from B:

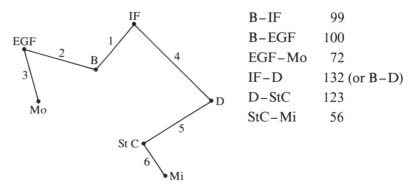

B–IF	99
B–EGF	100
EGF–Mo	72
IF–D	132 (or B–D)
D–StC	123
StC–Mi	56

7. (a) Bemidji is the city that is most central.

(b) $\max_u \min_v d(u, v) \le \min_u \max_v d(u, v)$.

(c) The most isolated city.

Section 4.8

1. (a)

Breadth-First Search

Label	Vertices	Unlabeled Adjacencies
0	O	A, C
1	A, C	B, F, D
2	B, D, F	E
3	E	—

3. (a)

Depth-First Search

Center	Next Center	Edge Labels	Center	Next Center	Edge Labels	Center	Next Center	Edge Labels
O	A	RY	C	O	RY	B	E	RB
A	B	RY	O	C	RB	E	D	RB
B	C	RY	C	A	RB	D	C	RB
C	D	RY	A	F	RB	C	B	RB
D	E	RY	F	D	RB	B	A	RB
E	F	RY	D	F	RB	A	O	RB
F	A	RY	F	E	RB			
A	C	RB	E	B	RB			

4. (a)

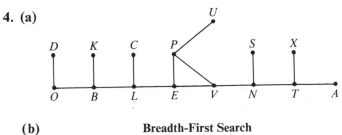

(b) **Breadth-First Search**

0	O	D, B	4	P, V	N, U
1	B, D	K, L	5	N, U	S, T
2	K, L	C, E	6	S, T	X, A
3	C, E	P, V	7	X, A	—

5. (b) Shortest path from O to M: $OCAHDVIBM$

Section 5.1

2. (a) Maximal flow from A to H is 8:

$$
\begin{array}{ll}
ACEJH & 1 \\
ACEJMQH & 1 \\
ACH & 6
\end{array}
$$

4.

To From	D_1	D_2	D_3	D_4	D_5	Units Shipped	Units Unshipped
P_1			700	300		1,000	0
P_2		900		600		1,500	0
P_3	600			200	1,000	1,800	200
Demands	600	900	700	1,100	1,000	4,300	200

Shipping cost: $C = 8,400$.

5. (a)

B		IF		D		StC		Mi		Mo		EGF		B
	99		132		123		56		204		72		100	

Total length = 786 miles.

6. (a)

Current Circuit	Length	Next Vertex	Insert Between	Length Increase
B, B	0	IF	B, B	$99 + 99 - 0$
B, IF, B	198	EGF	B, IF	$100 + 168 - 99$
B, EGF, IF, B	367	Mo	B, EGF	$102 + 72 - 100$
B, Mo, EGF, IF, B	441	D* (or StC)	IF, B	$132 + 132 - 99$
B, Mo, EGF, IF, D, B	606	StC	D, B	$132 + 123 - 132$
B, Mo, EGF, IF, D, StC, B	729	Mi	D, StC	$127 + 56 - 123$

Hamilton circuit: B, Mo, EGF, IF, D, Mi, StC, B.
Length 789 miles.

Section 5.2

2. (a) In the order in which the sets are listed, an SDR (system of distinct representatives) is given by {*e, c, d, b, a*}.

(c) No SDR exists because the 4 sets contain only 3 elements.

4.

Pitcher		*B*	*C*
Catcher	*E*		
First	*G*		
Second	*D*		
Third	*H*		
Shortstop	*J*		
Left		*C*	*F*
Center	*A*		
Right		*F*	*B*

Only 2 lineups are possible.

8. (b) {*a, y, c, v, e, x*}.

9. (b) {*x, y*}, {*a, b*}, {*u, v*}, {*c, d*}, {*e, w*}.

Section 5.3

1. (a)

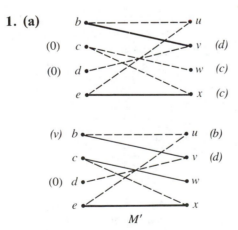

M'

{*c, w*} is an alternating chain of length 1. Hence adjoin {*c, w*} to *M* to obtain *M'*, as in the sketch.

Alternating chain of odd length: {*d, v*}, {*v, b*}, {*b, u*}. Delete {*v, b*} from *M'*; adjoin {*d, v*} and {*b, u*} to obtain *M''*.

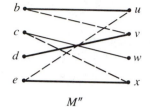

M''

Clearly M'' is a complete (and hence maximal) matching. An additional application of the algorithm produces a vertex cover $C = \{u, v, w, x\}$ with 4 vertices (and hence minimal).

3. (a)

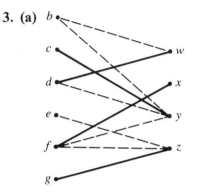

Choose as a minimal vertex cover the 4 vertices of Z: $K = \{w, x, y, z\}$. The set of edges $M = \{\{d, w\}, \{f, x\}, \{c, y\}, \{g, z\}\}$ is a maximal matching. (Other choices of M and K are available.)

Section 5.4

1. (i) (a) f: SA (3) SB (5) SC (2)
    AE (2) BE (3) CF (1)
    AF (1) BG (2) CG (1)
    ED (5) FD (2) GD (3)
        $v(f) = 10$

(b) Let $X = \{S\}$, $Y = \{A, B, C, E, F, G, D\}$.
    $C = \{X, Y\}$ is a cut of capacity $3 + 5 + 2$, the sum of the capacities of SA, SB, SC.

(c) Because $v(f) = c(C) = 10$, f is a maximal flow and C is a minimal cut.

(iii) (a) f: SA (5)
    AE (2)
    AF (3) SB (3)
    FB (1) BE (3 + 1)
    FD (2) ED (2 + 4)
        $v(f) = 8$

(b) Let $X = \{S\}$, $Y = \{A, B, E, F, D\}$.
    $C = \{X, Y\}$ is a cut of capacity 8.

(c) Because $v(f) = v(C) = 8$, f is a maximal flow and C is a minimal cut.

3. (a) If $C = \{X, Y\}$ is a cut, then S is in X and D is in Y. The other 4 vertices can be assigned independently to X or Y, so there are 16 possible cuts.

Section 5.5

1. (a)

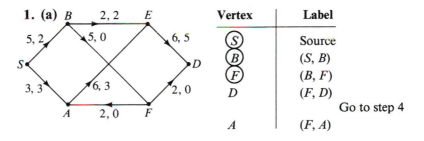

Vertex	Label
S	Source
B	(S, B)
F	(B, F)
D	(F, D)
	Go to step 4
A	(F, A)

$$
\begin{array}{cccc}
S & B & F & D \\
& 3 & 5 & 2
\end{array}
$$

Let $|\Delta f| = 2$, and let $f' = f + \Delta f$. $v(f_1) = 5 + 2 = 7$.

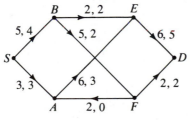

Vertex	Label
S	Source
B	(S, B)
F	(B, F)
A	(F, A)
E	(A, E)
	Go to step 4
D	(E, D)

$$
\begin{array}{cccccc}
S & B & F & A & E & D \\
& 1 & 3 & 2 & 3 & 1
\end{array}
$$

Let $|\Delta f| = 1$ and let $f'' = f' + \Delta f$. $v(f'') = 7 + 1 = 8$.

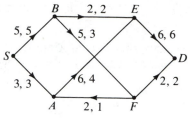

Vertex	Label
Ⓢ	Source

So $\{S\}$ and $\{A, B, E, F, D\}$ is a cut of the network with capacity $5 + 3$, equaling current flow. So this flow is maximal, and the cut is minimal.

3. The cut, $\{S, A, B, C, E, F, G\}$ and $\{H, J, K, D\}$, has a capacity of 17, so no flow can exceed 17, and any flow of 17 is maximal.

Index